Arsenic and Old Mustard: Chemical Problems in the Destruction of Old Arsenical and 'Mustard' Munitions

NATO ASI Series

Advanced Science Institute Series

A Series presenting the results of activities sponsored by the NATO Science Committee, which aims at the dissemination of advanced scientific and technological knowledge, with a view to strengthening links between scientific communities.

The Series is published by an international board of publishers in conjunction with the NATO Scientific Affairs Division

A Life Sciences	Plenum Publishing Corporation
B Physics	London and New York
C Mathematical and Physical Sciences	Kluwer Academic Publishers
D Behavioural and Social Sciences	Dordrecht, Boston and London
E Applied Sciences	
F Computer and Systems Sciences	Springer-Verlag
G Ecological Sciences	Berlin, Heidelberg, New York, London,
H Cell Biology	Paris and Tokyo
I Global Environment Change	

PARTNERSHIP SUB-SERIES

1. Disarmament Technologies	Kluwer Academic Publishers
2. Environment	Springer-Verlag / Kluwer Academic Publishers
3. High Technology	Kluwer Academic Publishers
4. Science and Technology Policy	Kluwer Academic Publishers
5. Computer Networking	Kluwer Academic Publishers

The Partnership Sub-Series incorporates activities undertaken in collaboration with NATO's Cooperation Partners, the countries of the CIS and Central and Eastern Europe, in Priority Areas of concern to those countries.

NATO-PCO-DATA BASE

The electronic index to the NATO ASI Series provides full bibliographical references (with keywords and/or abstracts) to about 50,000 contributions from international scientists published in all sections of the NATO ASI Series. Access to the NATO-PCO-DATA BASE is possible via a CD-ROM "NATO Science and Technology Disk" with user-friendly retrieval software in English, French, and German (©WTV GmbH and DATAWARE Technologies, Inc. 1989). The CD-ROM contains the AGARD Aerospace Database.

The CD-ROM can be ordered through any member of the Board of Publishers or through NATO-PCO, Overijse, Belgium.

1. Disarmament Technologies – Vol. 19

Arsenic and Old Mustard: Chemical Problems in the Destruction of Old Arsenical and 'Mustard' Munitions

edited by

Joseph F. Bunnett
University of California,
Santa Cruz, California, U.S.A.

and

Marian Mikołajczyk
Centre of Molecular and Macromolecular Studies,
Polish Academy of Sciences,
Łódź Poland

Kluwer Academic Publishers

Dordrecht / Boston / London

Published in cooperation with NATO Scientific Affairs Division

Proceedings of the NATO Advanced Research Workshop on
Chemical Problems Associated with Old Arsenical and 'Mustard' Munitions
Łódź, Poland
17-19 March 1996

A C.I.P. Catalogue record for this book is available from the Library of Congress.

ISBN 978-90-481-5069-4

Published by Kluwer Academic Publishers,
P.O. Box 17, 3300 AA Dordrecht, The Netherlands.

Sold and distributed in North, Central and South America
by Kluwer Academic Publishers,
101 Philip Drive, Norwell, MA 02061, U.S.A.

In all other countries, sold and distributed
by Kluwer Academic Publishers,
P.O. Box 322, 3300 AH Dordrecht, The Netherlands.

Printed on acid-free paper

CONTENTS

LECTURES

POSTER ABSTRACTS

GROUP DISCUSSION REPORTS

APPENDICES

PREFACE

Chemical warfare, involving substances often called "poison gases", was initiated by German forces in 1915. It was soon taken up by the Allies and was actively practiced until the end of World War I in 1918.[1] The original chemical warfare (CW) agents were truly gases (i.e., chlorine, phosgene and hydrogen cyanide), but liquid Yperite or 'mustard' caused the majority of World War I "gas" casualties. Yperite is a vesicant; it raises huge blisters on exposed skin. Its worst effects were however on persons who inhaled mists or aerosols of Yperite; the damage to delicate lung tissue was often irreversible. Some arsenic compounds were also used as chemical warfare agents. Most of them (Clark I, Adamsite, etc.) induced sneezing or vomiting and were intended mainly to disorient enemy troops. Lewisite, an arsenical vesicant, was developed during World War I and produced in large quantity by the Soviet Union in the Cold War era, but has seen little battlefield use.

Forces on both sides in World War II manufactured vast amounts of chemical munitions, but they were not used in the war. During the Cold War, the Soviet Union and the United States also armed themselves with prodigious amounts of chemical weapons, but again they were not used. Problems of destruction of those huge stockpiles of chemical munitions in Russia and the United States have justifiably received much attention, in both their technological and political aspects. But the problems in destroying "non-stockpile" munitions are even greater.

Those "non-stockpile" munitions lie buried on old battlefields or military posts or sunken in estuaries or the sea. It is estimated that in Europe alone over ten million lost or abandoned World War I chemical munitions, many of them still charged with agent, are yet to be retrieved and destroyed. Many European nations in which that war was fought have departments to destroy the old munitions that are frequently found by citizens as they engage in activities such as construction, farming or fishing.

There are major problems also in China, where Japanese forces abandoned chemical munitions in 1945. In the United States, the U.S. Army has surveyed its substantial non-stockpile chemical munitions problem.[1] One aspect of it came to light in 1993, when abandoned chemical munitions and precursor chemicals were found buried in Spring Valley, Washington, D.C., a district of expensive homes. During World War I it was a chemical weapons development site.

Chemical Problems

A discovered old munition might be of either high explosive or chemical type. Weapons destroyers must determine which it is, and if chemical what agent it contains. The internal structure of a munition can be revealed and often indications of its chemical content obtained by ultrasound and X-ray analysis. For chemical identification, neutron activation analysis often gives firm indications.

Inasmuch as Yperite is a liquid under most environmental conditions (it freezes at 14.4 °C), one might anticipate that an opened Yperite munition could simply be drained and the resulting liquid destroyed by incineration or other technology. It has however been found that containers of Yperite, including Yperite-filled munitions, that have stood for decades often contain gelled or solid material formed by chemical reactions during those many years. Such deposits impede the emptying of munitions and the necessary ultimate cleansing of them so that the metal can safely be disposed of as scrap. The chemical nature of the deposits, of gases that sometimes form as well, and of the reactions that formed them are incompletely understood. It is unlikely that all the deposits are of the same chemical character. Understanding those deposits, and devising technologies to deal with them, are serious problems in chemical weapons destruction.

Destruction of arsenical agents presents problems of a different sort. The essential difficulties are that many arsenic compounds are very toxic and that any treatment of an arsenic-containing chemical warfare agent unavoidably furnishes one or more arsenic-comtaining products. What arsenic-containing products are most acceptable, and what disposal of them is safe?

Genesis of this NATO Advanced Research Workshop

As a step toward solution of the mentioned problems, the Committee on Chemical Weapons Destruction Technologies of the International Union of Pure and Applied Chemistry (IUPAC), in cooperation with chemical weapons experts from the United Kingdom and the United States, sought assistance from the NATO Scientific Affairs Division in staging a workshop for examination of them. NATO agreed to provide support, and a NATO Advanced Research Workshop entitled "Chemical Problems Associated with Old 'Mustard' and Arsenical Munitions" was organized. The co-chairmen of this NATO ARW are co–editors of this book; both are members of the mentioned IUPAC committee.

An objective of the NATO ARW organizers was to bring technologists who work at destruction of chemical munitions into intellectual contact with chemists doing fundamental research in relevant areas. It was believed that concepts developed in fundamental research would be useful to practitioners of chemical demilitarization and that the research chemists, once they knew of practical problems, would see ways in which fundamental chemical principles might be applied toward solution of them.

In this volume the proceedings of the NATO ARW are published. Presented are texts of the lectures by key speakers (with one exception), and abstracts of posters presented. Reports of three discussion groups, concerning three major problem areas, present consensus views.

Joseph F. Bunnett, NATO ARW Co-chairman
Marian Mikolajczyk, NATO ARW Co-chairman

References

1. Haber, L.F. (1986) *The Poisonous Cloud: Chemical Warfare in the First World War*, Oxford University Press, New York

2. Program Manager for Non-Stockpile Chemical Materiel (1993) *Survey and Analysis Report*, U.S. Army Chemical Materiel Destruction Agency

ACKNOWLEDGMENTS

Major support for this Advanced Research Workshop was provided by the North Atlantic Treaty Organization through its Scientific Affairs Division, Programme on Disarmament Technologies. A favorable recommendation by the Advisory Panel on Disarmament Technologies was and is appreciated.

Important support in Poland was furnished by the State Committee of Scientific Research (KBN), the City Council of the City of Lodz, and the Lodz Branch of the Polish Academy of Sciences. The ARW was hosted by the Centre of Molecular and Macromolecular Studies of the Polish Academy of Sciences, Lodz, Poland.

In its intellectual dimensions, the NATO ARW was co-sponsored by the Committee on Chemical Weapons Destruction Technologies of the International Union of Pure and Applied Chemistry (IUPAC). The work of this Committee was supported importantly by the John D. and Catherine T. MacArthur Foundation (U.S.) and by the Consiglio Nazionale delle Ricerche d'Italia, and significantly by IUPAC, the UK Chemical and Biological Defence Establishment, the U.S. Army European Research Office, and the University of California, Santa Cruz.

Organization of this NATO ARW was facilitated by the advice of Mrs. Nancy Schulte of the NATO Scientific Affairs Division, the counsel of members of the Organizing Committee, and the work of Kim Lisa Jackson (UC Santa Cruz) and (in Lodz) of Drs. Piotr Kiełbasiński (NATO ARW Secretary), Wanda Midura, Bogdan Bujnicki, Piotr Łyżw and other members of the Local Organizing Committee.

Structures and Common Names of Prominent Arsenical and 'Mustard' Agents

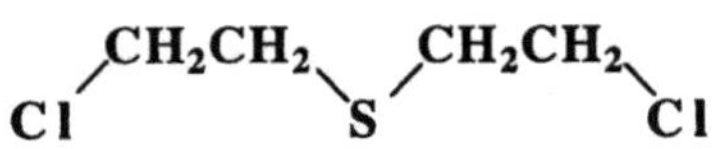

Yperite or "Mustard"

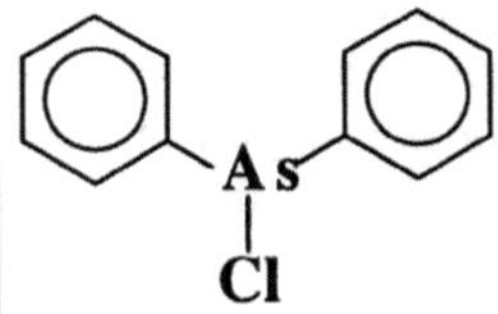

Clark I

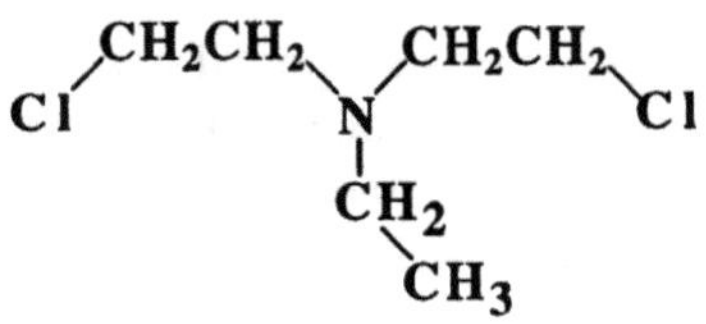

A "Nitrogen Mustard"

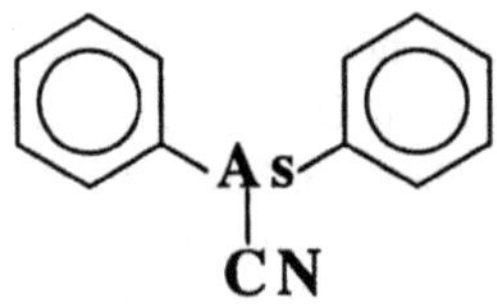

Clark II

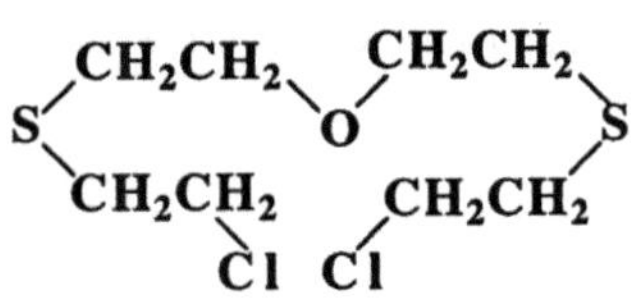

Agent T

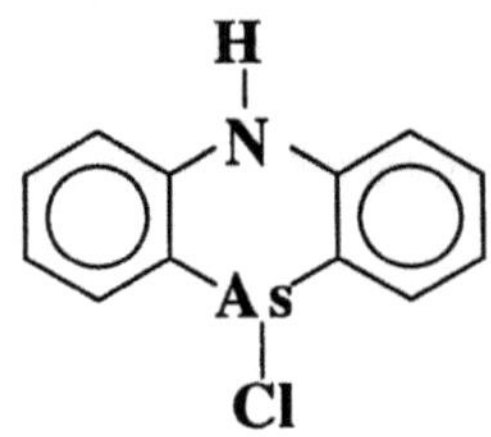

Adamsite

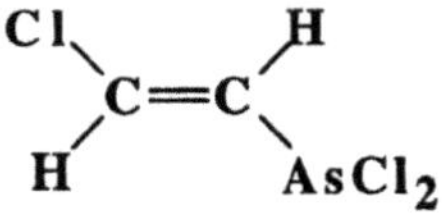

Lewisite

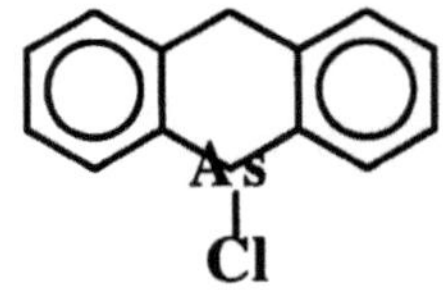

Excelsior

IUPAC NAMES OF PROMINENT CHEMICAL WARFARE AGENTS

ARSENICAL AGENTS

Adamsite: 10-chloro-5,10-dihydrophenarsazine

Clark I: diphenylarsinous chloride

Clark II: cyanodiphenylarsine

Phenyldichloroarsine ($PhAsCl_2$): phenylarsonous dichloride

Lewisite: 2-chloroethenylarsonous dichloride

'MUSTARD'-TYPE AGENTS

Yperite: bis(2-chloroethyl) sulfide

Agent T: bis[2-(2-chloroethylthio)ethyl] ether

A 'nitrogen mustard' (from page xii): bis(2-chloroethyl)ethylamine

MEMBERS OF THE ORGANIZING COMMITTEE

Joseph F. Bunnett (Co-Director)
Professor of Chemistry Emeritus
University of California, Santa Cruz, CA 95064, USA

Marian Mikolajczyk (Co-Director)
Professor of Chemistry and Director,
Centre of Molecular and Macromolecular Studies
Polish Academy of Sciences,90-363 Lodz, Poland

Irina P. Beletskaya
Professor of Chemistry
Moscow State University, Moscow, Russia

Robert W. Shaw
Associate Director, Division of Chemical and Biological Sciences
U.S. Army Research Office, Research Triangle Park, NC, USA

Richard J. Soilleux
Head of Safety Services
Chemical and Biological Defence Establishment, Porton Down,
Salisbury, UK

GLOSSARY OF CODES FOR SELECTED ARSENICAL AND 'MUSTARD' AGENTS

COMPILED BY BENJAMIN C. GARRETT[1]
Battelle Memorial Institute,
Arlington, VA 22202-4172, USA

CODE	EXPLANATION	NATION
Adamsite	10-chloro-5,10-dihydrophenarsazine	US
Arsinöl	mix of diphenylchloroarsine (35%), phenyldichloroarsine (50%), triphenylarsine (10%) and arsenic(III) chloride (5%)	Germany
Azin	Adamsite	Germany
BB	'sulfur mustard'	UK
C 6-Base	'nitrogen mustard'	Germany
C 6-Salz	'nitrogen mustard' hydrochloride	Germany
CBR	mix of phosgene (50%) + arsenic (III) chloride	UK
CD	diphenylcyanoarsine	UK
CDA	diphenylchloroarsine	US
Clark I	diphenylchloroarsine	Germany
Clark II	diphenylcyanoarsine	Germany
DA	diphenylchloroarsine	US
DC	diphenylcyanoarsine	US
DESA	'sulfur mustard' via sulfur monochloride process	UK
DESAV	viscous DESA	UK
Dick	diphenylchloroarsine	Germany
DJ	phenyldichloroarsine	US
D-Lost	'sulfur mustard' produced from ethylene and sulfur dichloride	Germany

[1] Compiled from published information. For an extensive listing of codes for World War II-era CW agents, see Stock, T. and Lohs, Kh. (1997) "Old chemical munitions and warfare agents: detoxification and degradation", *The Challenge of Old Chemical Munitions and Toxic Armament Wastes*, (Oxford University Press: Oxford, pp. 35-52. See also the paper of Martens (this volume, page 33).

DM	Adamsite	US
Doppellost	'sulfur mustard'	Germany
Dora	cyanodiphenylarsine	Germany
EBA	HN-1	UK
Ethyl-Dick	ethyldichloroarsine	Germany
Ethyl-S	HN-1	Germany
ED	ethyldichloroarsine	US
Excelsior	5-chloro-5,10-dihydroarsacridine	Germany
Fluorlost	bis-2-fluoroethyl sulfide	Germany
H	'sulfur mustard'	UK, US
HBV,HBDV	'sulfur mustard' + chlorobenzene made viscous by addition of chlorinated rubber	UK
HD	(distilled) 'sulfur mustard'	UK, US
HL	mix of 'sulfur mustard' (63%) + Lewisite (37%)	US
HLV	viscous HL	US
HN-1	bis(2-chloroethyl)ethylamine	US
HN-2	bis(2-chloroethyl)methylamine	US
HN-3	tris(2-chloroethyl)amine	US
HQ	mix of 'sulfur mustard' + Q	US
HS	'sulfur mustard'	UK, US
HT	mix of 'sulfur mustard' (60%) + T (40%)	US
JBR	mix of hydrogen cyanide (50%), arsenic(III) chloride (25%) and chloroform (25%)	UK
L-1	2-chlorovinyldichloroarsine	US
L-2	bis(2-chlorovinyl)chloroarsine	US
L-3	tris(2-chlorovinyl)arsine	US
Lewisite, L	L-1 and/or L-2 and/or L-3	US
Lost	'sulfur mustard'	Germany
M1	2-chlorovinyldichloroarsine	US
M2	bis(2-chlorovinyl)chloroarsine	US
M3	tris(2-chlorovinyl)arsine	US
M 30	mix of Adamsite, cellulose nitrate, urea, diethyl phthalate and magnesium oxide	Germany
M 49	mix of Adamsite, cellulose nitrate, urea, and diethyl phthalate	Germany
Manganite	mix of hydrogen cyanide (50%) and arsenic(III) chloride (50%)	France
Marsite	arsenic (III) chloride	France
MBA	bis(2-chloroethyl)methylamine (HN-2)	UK

MD	methyldichloroarsine	US
Medikus	methyldichloroarsine	Germany
Methyl-Dick	methyldichloroarsine	Germany
Mustard	'sulfur mustard'	
Nitrogen mustard	general term for substituted alkyl(2-chloroethyl)amines	US
NL, N-Lost	tris(2-chloroethyl)amine (HN-3)	Germany
O, Oxol-Lost	'sulfur mustard' from thiodiglycol	Germany
OB	bis[2-(2-chloroethylthio)ethyl] ether	Germany
Oxygen mustard	bis[2-(2-chloroethylthio)ethyl] ether	US
PD	phenyldichloroarsine	US
Phenyl-Dick	phenyldichloroarsine	Germany
Pfiffikus	phenyldichloroarsine	Germany
Propyllost	2-chloroethyl 2-chloropropyl sulfide	Germany
Q	1,2-bis(2-chloroethylthio)ethane	UK, US
R-15[2]	Adamsite	USSR
R-43A[3]	Lewisite	USSR
R-74[4]	'sulfur mustard'	USSR
Red no. 1	diphenylcyanoarsine	Japan
RK-7[5]	mix of 'sulfur mustard' and Lewisite (ostensibly, 50%/50%)	USSR
S	bis(2-chloroethyl)methylamine	UK
Sengas	'sulfur mustard'	Germany
Sesqui mustard	1,2-bis(2-chloroethylthio)ethane	US
Sulfur mustard	bis(2-chloroethyl) sulfide	
S-yperit	'sulfur mustard'	Germany
T	bis[2-(2-chloroethylthio)ethyl] ether	UK, US
TBA	tris(2-chloroethyl)amine	Germany
T-9	tris(2-chloroethyl)amine	Germany
Trilon 300	arsenic hydride	Germany
UP	tris(2-chloroethyl)amine	Germany

2 in Cyrillic: P-15

3 in Cyrillic: P-43A

4 in Cyrillic: P-74

5 in Cyrillic: PK-7

Vincennite	mix of arsenic(III) chloride (30%), stannic chloride (15%), hydrogen cyanide (50%), and chloroform (5%)	France
VIR	'sulfur mustard'	USSR
Vitrite	mix of cyanogen chloride (70%) and arsenic(III) chloride (30%)	France
VN	Vincennite	UK
VRK-7[6]	viscous mix, 'sulfur mustard' & Lewisite	USSR
Winter-Lost	mix of 'sulfur mustard' (50%) and phenyldichloroarsine (50%) plus freezing point depression additives[7]	Germany
White no. 1	arsenic(III) chloride	Japan
Y1	'sulfur mustard'	UK
Y2	viscous 'sulfur mustard'	UK
Y3	HT	UK
Y4	viscous HT	UK
Y5	'sulfur mustard' prepared from sulfur dichloride and ethene	UK
Y5A	Y5 in chlorobenzene	UK
Y5B	Y5 in chlorobenzene	UK
Y5C	Y5 in chlorobenzene	UK
Y6	viscous Y5	UK
Y7	Lewisite	UK
Y9, YN	HL	UK
Y10	HLV	UK
Y13	Y5	UK
Y14	HBV	UK
Y15	'sulfur mustard' from sulfur monochloride process	UK
Y16	viscous Y15	UK
Yellow no. 1	'sulfur mustard'	Japan
Yellow no. 2	Lewisite	Japan
Yperite	'sulfur mustard'	France
Zählost	viscous 'sulfur mustard'	Germany
Zähyperit	viscous 'sulfur mustard'	Germany
ZL	viscous 'sulfur mustard'	Germany

---

[6] in Cyrillic: BPK-7

[7] Such as arsinöl, chlorobenzene or anthracene.

THE PROBLEM OF OLD CHEMICAL WEAPONS WHICH CONTAIN "MUSTARD GAS" OR ORGANOARSENIC COMPOUNDS: AN OVERVIEW

RON G. MANLEY
Head of Chemical Demilitarisation Branch
Provisional Technical Secretariat of the Preparatory Commission for the Organisation for the Prohibition of Chemical Weapons

1. Introduction

In recent years considerable publicity has been given to the problem of disposing of the large stockpiles of chemical weapons held by the Russian Federation (ca. 40,000 tonnes) and the United States of America (ca. 31,000 tonnes). Whilst the disposal of these large stockpiles of weapons in a safe and environmentally acceptable manner is of great importance, it has nevertheless tended to overshadow the equally important problem of disposing of the large numbers of old and abandoned chemical weapons scattered throughout the world. Whilst, unlike the Russian and U.S. stockpiles, these old chemical weapons generally no longer pose any significant military threat, they may still pose a potential threat to the people and the environment in the areas where they are located. This paper attempts to provide an overview of the problem posed by these old and/or abandoned chemical weapons and the practicalities associated with their recovery and safe disposal. Particular emphasis is placed on the problems associated with old chemical weapons containing either thickened or polymerised "mustard gas" and/or organic arsenicals, as these have been shown to pose some of the most difficult disposal problems. More detailed examples of some of these problems are provided in the papers by other authors which follow this introductory overview.

2. Brief History

There are many recorded incidents throughout history where chemicals have been used in an attempt to gain a victory or advantage during a war. The Spartans, for example, are reported to have used arsenic smoke during the Peloponnesian War (431-404 BC) [1] and the first recorded use of "Greek Fire", a mixture of petroleum, pitch, sulphur and various resins, was by the Byzantine Greeks at the siege of Constantinople in 673 AD.[1] The history of modern chemical warfare, however, is generally accepted to have begun with the release, by Germany, on 22 April 1915 of 168 tonnes of chlorine gas from around 6,000 cylinders at the battle of Ypres. [2] This first large-scale use of gas subsequently

J.F. Bunnett and M. Mikołajczyk (eds.), Arsenic and Old Mustard: Chemical Problems in the Destruction of Old Arsenical and 'Mustard' Munitions, 1–16.

led to the rapid development of the necessary technology and the extensive use of chemical weapons by both sides during the remaining years of World War I.

During the period 1915 to 1918 many thousands of chemicals were screened for their potential usefulness as chemical warfare agents and more than 40 different compounds were actually weaponised in a wide range of munitions. They were either used separately or as mixtures containing sometimes up to as many as five different components. Chemicals or combinations of chemicals were selected to produce a broad spectrum of effects ranging from simple harassment (irritant and choking gases) to severe incapacitation or death (blood agents and vesicants) Some examples of the principal chemicals used are given in table 1.

TABLE 1. Examples of chemicals used on a large scale in World War I as chemical agents.

Chemical	Code Name	Toxic Effect
Bromoacetone	BA, Martonite, B–Stoff	eye and lung irritant
Ethyl iodoacetate	SK	eye and lung irritant
Chlorine	Bertholite	choking agent
Trichloronitromethane	Chloropicrin, PS, Klop	choking agent
Phosgene	CG, Zusatz, Collongite	choking agent
Hydrogen cyanide	AC, Forestite	blood agent
Phenyldichloroarsine	PD	vomiting agent
Diphenylchloroarsine	DA, Clark I	vomiting agent
Bis–(2chloroethyl) sulphide	"Mustard gas"	vesicant

The most significant step, however, was the introduction of "mustard gas" [bis (2-chloroethyl) sulphide] at the second battle of Ypres in 1917, and its escalating use from this point until the end of the war. "Mustard gas" produced large numbers of casualties on both sides and rapidly became known as the "king of war gases" and, despite the development of chemical agents with far greater toxicity over the years, is still regarded by many experts in the field to continue to warrant this title even today.

Records show that gas attacks were a routine occurrence from 1915 until the end of the war in November 1918. [3] The total tonnage of chemical agents used during WWI is quoted as 124,200 tonnes, and the combined total number of artillery shells containing chemical agents fired by all belligerents during the period 1915 to 1918 is quoted as 66 million. [4] The same source also estimates that approximately 87% of all chemical agents were delivered by this means. An indication of the scale of some of these attacks can be obtained from the first recorded use of "mustard gas" in July 1917. During the single night of 12-13 July the Germans fired 50,000 shells containing "mustard gas" into the Ypres Salient [5], and during the following ten days more than one million shells containing this chemical agent were used by the Germans. [6] Other records [5] suggest that between 20-30% of the munitions used during WWI failed to function. Assuming that these records are approximately correct, simple arithmetic would suggest that somewhere between 13 million and 20 million chemical rounds may have remained scattered across the battlefield at the end of the war. Whilst many of these rounds were recovered and destroyed during the 1920s, others have remained buried on the old battlefields. Farmers in the Flanders region continue to uncover these unexploded munitions on a routine basis more than 75 years after the end of the war.

In the period between WWI and WWII most of the major powers continued their research into and development of chemical weapons. The experiences of WWI led them to believe that chemical weapons would play a major part in any future wars. It is therefore understandable that, with the onset of WWII, the major powers began to produce and stockpile these weapons in very large numbers and continued to do so throughout the war. The total tonnage of chemical agents produced exceeded that produced during WWI. For a variety of reasons, however, chemical weapons were not widely used during WWII. The major exceptions were the large-scale use by the Italians in Ethiopia during the period 1935-38 and by the Japanese in Manchuria, China, during the years 1937-43. Chemical weapons were, however, transported and stockpiled in each major war theatre, including the Western Front, Eastern Front, Middle East, North Africa, Asia and the Pacific. At the end of the war stocks of chemical weapons were therefore present in many countries. The precise fate of many of these stockpiles is uncertain. Some were shipped back to the countries that produced them, some were dumped at sea in the region where they had been stockpiled, and others were simply buried at or near the stockpile sites. The most notable example of the last option are the estimated two million chemical munitions reported to have been buried and abandoned by the Japanese at various sites in China.[7]

There have been a number of incidents over the years when caches of these munitions or bulk tanks of chemical agent were accidentally rediscovered, and had to be recovered and destroyed. For example, in West Java, Indonesia, a number of underground bulk tanks containing around 45 tonnes of "mustard gas" which had been manufactured during WWII by the Government of the then Netherlands East Indies were rediscovered in 1978. This material was recovered and safely destroyed by Dutch experts using incineration during 1979. [8] In 1987 approximately 400 chemical munitions, many of them fuzed, were discovered during the installation of a new water main at an old ammunition depot in Hampshire in the United Kingdom. [9] The United States undertook a recovery operation of chemical munitions uncovered in the Solomon Islands in 1988-89. The recovered chemical weapons (c.a. 100) were transported to the US destruction facility at Johnston Atoll for disposal in 1991. [10] In 1993 there was the well reported discovery of old chemical munitions in the Spring Valley area of Washington, DC. [11] It is probable that more of these discoveries will be made in the future as land is redeveloped for other uses. The exact number of countries which have potential old and/or abandoned chemical weapons as a result of activities over the last 75 years is again not known, but is thought to be somewhere between 20 and 50.

After WWII a number of countries continued to maintain active chemical weapons programmes. In particular the United States of America and the Soviet Union produced and maintained large stockpiles of modern chemical weapons containing the highly toxic nerve agents. Other countries, notably Iraq, have also developed and stockpiled chemical weapons.

3. WWI and WWII Chemical Agents

3.1 ARSENICALS

As noted in earlier paragraphs, many chemicals were developed and used as chemical agents during WWI. In the years following the war this development continued up to and throughout WWII. The first chemical agents were disseminated in vapour form and aimed at attacking the lungs. However, these were rapidly countered by the development of gas masks. This led to the development of the organoarsenic vomiting or harassing agents. These were generally solids which had to be disseminated either as solutions or by the use of high explosives. In the former case the solution was disseminated in the form of droplets which rapidly evaporated to produce the chemical agent in the form of fine particles. In the latter case the explosive charge was used to produce the cloud of fine particles. These fine particles were capable of defeating the early gas mask filters, and thus caused the soldiers to choke or vomit and, as a result, forced them to remove their gas masks. The development of particulate filters eventually reduced the effectiveness of these agents against protected troops. The next step was the development of chemical agents which attacked the skin as well as the lungs.

TABLE 2. Arsenic based chemical warfare agents

Compound	Code Name	Remarks
Arsine	SA	Toxic gas known as "Mithrite" in France, developed but not used during WWII.
Diphenylchloroarsine	DA	Vomiting agent developed and used in WWI, also stockpiled during WWII.
Diphenylcyanoarsine	DC, red gas	Vomiting agent, more toxic than DA. Developed in 1918. Stockpiled by several countries and also used by Japan in China during WWII.
Phenyldibromoarsine		Minor WWI agent
10–chloro–5, 10–dihydrophenarsine	Adamsite (DM)	Vomiting agent develped after WWI, stockpiled during WWII.
Phenyldichloroarsine	PD	Vomiting and vesicant properties. WWI agent also stockpiled during WWII.
Ethyldichloroarsine	ED	WWI vesicant agent.
Methyldichloroarsine	MD	Lung irritant with vesicant properties. Limited use during WWI.
Ethyldibromoarsine		Minor WWI agent used as a "mustard gas" additive.
2–Chlorovinyldichloroarsine	Lewisite (L)	Vesicant and systemic poison developed in 1918. Stockpiled during WWII and used by Japan in China.

Although the most notable of these was "mustard gas" , a number of organoarsenical vesicant agents were also developed. These arsenical vesicants, such as ethyldichloroarsine or Lewisite, cause immediate excruciating pain upon contact with the eyes and a stinging pain on the skin. The pattern of lung attack is similar to that of "mustard gas" , but the systemic poisoning effect is much greater. A list of the principal arsenic containing compounds either developed and used during WWI or developed during the period immediately after the war is provided in table 2. Examples of some of the mixtures developed and weaponised which contained arsenic compounds are provided in table 3

3.2 "MUSTARD GAS"

Pure "mustard gas" solidifies at 14°C. This poses difficulties for use in cold weather conditions and several solutions to this problem were developed over the years. During WWII, for example, the British developed and produced on a large scale a mixture known as HT which had a freezing point of -1 °C. This was a mixture of 60% "mustard gas" and 40% bis (2-chloroethylthioethyl) ether (T). The vesicant properties of T were in fact 3 to 4 times greater than those of "mustard gas" . A second solution to the problem was to depress the freezing point by adding one of the arsenical vesicants such as Lewisite, phenyldichloroarsine or ethyldibromoarsine. This latter approach was favoured by the Germans, the Japanese and the Soviets. In order to improve its dissemination and persistence, "mustard gas" and its low freezing point mixtures were often thickened by the addition of polymeric materials. Initially "Alloprene", a chlorinated rubber, was used for this purpose but this was later replaced by poly[methyl methacrylate](PMMA).

TABLE 3. Some examples of chemical mixtures deployed in WWI munitions which contained arsenic compounds

Code Name	Origin	Chemicals	Percentage
Manganite	France	Hydrogen cyanide Arsenic trichloride	50 50
Vincennite	France	Hydrogen cyanide Stannic chloride Arsenic trichloride Chloroform	50 15 30 5
CBR	Britain	Phosgene Arsenic trichloride	50 50
JBR	Britain	Hydrogen cyanide Arsenic trichloride Chloroform	50 25 25
Grunkreuz 2	Germany	Phosgene Trichloromethyl chloroformate Diphenylchloroarisne	60 30 10
Gelbkreuz 1	Germany	Ethyldichloroarsine Ethyldibromoarsine "Mustard gas"	40 40 20

Even in munitions containing relatively pure "mustard gas" the material will begin to react over a period of time to form polymers and other complex sulphur containing products. The polysulphide polymer formed is insoluble in most solvents, acts like a sponge and absorbs much of the remaining "mustard gas" and other reaction products, many of which are also highly vesicant. The decommissioning of old munitions containing "mustard gas" or "mustard gas" mixtures is therefore by no means a simple process.

3.3 PRODUCTION AND STOCKPILING

As stated earlier, the Germans first used "mustard gas" in 1917. The French and the British had to build special plants to produce "mustard gas", and did not have supplies available until well into 1918. As a result of its use during the closing years of WWI, many countries appreciated the military potential of "mustard gas", and decided to acquire their own stocks. During WWII "mustard gas" was reportedly manufactured in significant quantities by the USA, the UK, France, Canada, Poland, the USSR, Hungary, Italy, Japan, and Germany. [12]

TABLE 4. Countries recorded as having produced and stockpiled arsenic based CW agents at some time during the period 1930–1945

	Lewisite	**Diphenyl chloroarsine**	**Diphenyl cyanoarsine**	**Adamsite**
France	yes			yes
Germany		yes		yes
Italy	yes	yes		yes
Japan	yes	yes	yes	
UK	some	yes	some	yes
USA	yes			yes
USSR	yes			yes

The development and production of the arsenical agents also continued, albeit on a smaller scale. Two chemicals in particular saw considerable further development. These were Lewisite (2-Chlorovinyldichloroarsine), a powerful vesicant and systemic poison and Adamsite (10-chloro-5,10-dihydrophenarsine), a potent harassing agent. Lewisite, named after its American discoverer, Dr Winford L Lewis, was being manufactured by the end of WWI, but the war ended before it could be deployed on the battlefield. Early tests suggested that it had a great potential as a chemical agent and therefore considerable effort was devoted to its development and production by a number of countries during WWII. The chemical agent properties of Adamsite are very similar to those of diphenylchloroarsine and diphenylcyanoarsine. However, it was much easier to produce and it also had the added advantage that it could readily be dispersed as a particulate smoke from thermal generators and smoke candles. An indication of which countries produced and stockpiled arsenic based chemical agents during WWII is given in table 4. It should be noted, however, that this list is by no means exhaustive.

The United States had production plants for Lewisite at its arsenals at Huntsville, Pine Bluff, and Rocky Mountain. Around 20,000 tonnes had been produced by the time the plants were closed in 1943 .[11] Much of this material was disposed of by dumping at sea after the war. [11] Its "mustard gas" production during the WWII period was approximately 80,000 tonnes. [10]

Between 1938 and 1942 the United Kingdom established chemical weapons production facilities in Randal, Valley, Springfields, Wade, Rocksavage, Millhouse and Roydmills and a research facility at Sutton Oak that were involved in the production and storage of "mustard gas" , phosgene, bromobenzyl cyanide, Lewisite, Adamsite, methyldichloroarsine and ethyl iodoacetate. In total approximately 60,000 tonnes of toxic agents were produced during WWII, of which about 40,000 tonnes were 'mustard' agents. [14] Much of the "mustard gas" was thickened to increase its persistence. Some of this material, particularly the "mustard gas" , was stored in bulk, while the remainder was filled into munitions. At the end of the war approximately 7,700 tonnes of "mustard gas" remained in bulk storage in the UK. [15] This material was subsequently destroyed by incineration.

Figures for German production during WWII of "mustard gas" and arsenicals were: "mustard gas" (25,000 tonnes), Adamsite (3,900 tonnes), diphenylchloroarsine (1,500 tonnes), diphenylcyanoarsine (100 tonnes) and arsenic oil - a mixture of phenyldichloroarsine (50%), diphenylchloroarsine (35%), arsenic (III) chloride (5%) and triphenylarsine (5%) - (7500 tonnes). [16]

A number of sources suggest that the quantity of "mustard gas" and arsenical agents produced by the USSR during WWII and after was very large, but it is difficult to confirm the accuracy of these figures for the reason that, as far as the author is aware, no official information has ever been published. One recent source gives the following figures for chemical agent production in the USSR for the period 1941 to 1945: "mustard gas" (76,700 tonnes), Lewisite (20,200 tonnes), and Adamsite (6,100 tonnes). Hydrogen cyanide (11,100 tonnes) and phosgene (8,300 tonnes) [17] The fate of all this material is unknown. As was the case with the other countries which produced chemical agents during WWII, much of it will have been filled into munitions. The remainder will have been stored in bulk. One source states that around 8,000 tonnes of Adamsite were buried near Shikhany in the Saratov region. [18]. The published figure for the current stockpile of "mustard gas" and Lewisite held by the Russian Federation is Lewisite (6,625 tonnes) "mustard gas" (690) and Lewisite/"mustard gas" mixture (210 tonnes) stored at Kembarka, Udmurtia region and at Gorny, Saratov region. [19] The period of manufacture of this material is, however, not quoted and therefore it may be post WWII material or a mixture of both.

Many difficulties arise in assessing how much and which CW agents Japan produced. This is due to a number of factors such as conflicting figures, including questionable identification of agents, and uncertainty as to whether or not weights quoted are for bulk agents or filled munitions. The following figures should therefore be treated with some caution. "Mustard gas" and Lewisite (c.a. 5,000 tonnes), diphenylcyanoarsine (ca. 2,000 tonnes), chloropicrin (ca. 1,000 tonnes), and phosgene (ca. 1.000 tonnes). [20] Much of this material was weaponised, and some of it was shipped to China for possible use by Japanese forces there.

4. WWI and WWII Chemical Munitions

Although the initial gas attacks were carried out using cylinders, the development of alternative means of delivery such as artillery shell and mortars soon commenced. In 1916 the British introduced the 4 inch Stokes mortar which contained six pounds of agent (2.7 kg) and had a range of up to approximately 1000 metres. This was followed by the Livens projector, named after its inventor Captain William H Livens. This device could fire a thin cased bomb approximately 200 mm in diameter and 500 mm long and filled with 13.6 kg of chemical for distances of up to nearly a mile. Three thousand of these projectors were used by the British at the beginning of the battle of Arras to fire almost 50 tonnes of phosgene in one day.[12] The solid arsenicals could be and were dispersed by the use of high explosives in artillery shells or mortars. An alternative was to use them in solution with HE. However, this means of dissemination was never found to be very satisfactory, and smoke candles or generators were considered to be more satisfactory. The basic candle dating from 1920-22 was replaced by a new design during 1941, the 9 lb M2 candle which contained 0.9 kg of DM. Some 92,000 were produced by the Edgewood arsenal during the war [13]. By end of the war a range of chemical munitions of all calibres had been developed and introduced into the battlefield by both sides. These included smoke candles, smoke generators, mines, mortars and artillery shells.

As stated earlier, although there was no large-scale use of chemical weapons during WWII, large numbers of chemical munitions were produced and filled. The United States, for example, acquired and filled with "mustard gas" approximately 0.5 million 4.2 inch mortars; 1.3 million 75 mm, 2 million 105 mm and 1 million 155 mm artillery shells; 0.6 million 115 pound bombs and 0.5 million 100 lb bombs; plus spray tanks and mines.[13] The majority of these munitions were filled with unthickened "mustard gas".

Germany produced a range of 'mustard' filled artillery shells, mortar shells, 250 and 500 kg bombs, rockets and spray tanks. They favoured using "mustard gas" in conjunction with other chemicals such as arsenic oil - a mixture of phenyldichloroarsine (50%), diphenylchloroarsine (35%), arsenic (III) chloride (5%) and triphenylarsine (5%) and diphenylchloroarsine. The Germans also used DM as a filling in base ejection shells, HE shells, aerial bombs and smoke candles. They also used a diphenylchloroarsine solution as a filling for rockets. After the war most of the German filled chemical munitions (130-140,000 tonnes) were disposed of by the Allies by sea dumping in the Skagerrak at a depth of 600 metres.[21]

Of the 40,000 tonnes of "mustard gas" produced by the United Kingdom, 23,000 tonnes were used to fill aerial bombs and 8,000 tonnes to fill more than 3 million shells, rockets and mortar bombs. [21] A large proportion of the UK stockpile of filled chemical munitions was dumped in deep water, inside scuttled ships, off the West coasts of Scotland and Ireland.

The Japanese also used 'mustard' for filling shells and bombs, primarily as a 50:50 mixture with Lewisite to lower the freezing point. Examples of Japanese chemical munitions are Artillery shells, 150 mm and 105 mm and Mortar rounds 90 mm, filled with either "mustard gas" /Lewisite mixture or diphenylcyanoarsine, and Artillery shells 75 mm filled with either phosgene or diphenylcyanoarsine. In addition there were 15 kg and 60 kg aerial bombs, toxic smoke candles and canisters filled with diphenyl-cyanoarsine. [12] Japan is estimated to have produced a total of around 1.7 million CW bombs and shells, and 5.7 million chemical candles and grenades. One source quotes the total amount of chemical munitions produced by Japan, regardless of agent fill, as approximately 7.4 million.[20] As already stated in an earlier paragraph, Chinese sources estimate that approximately 2 million of these rounds were abandoned in Northern China.[7]

From the above paragraphs it can be seen that all of the major powers produced large numbers of chemical munitions during WWII. The majority of these were filled with "mustard gas" , either on its own or in combination with other chemical agents and/or modifying chemicals. Although the total numbers are much smaller, many were also filled with arsenical agents. Whilst detailed records are often no longer available, those that are indicate that chemical weapons were shipped and stockpiled in every major war theatre. The inventory of U.S. chemical weapons stockpiled in Australia during early 1943, given in Table 5, is just one example of these overseas stockpiles. This stockpile increased significantly as the Pacific campaign progressed. [6] Records suggest that most of this material was subsequently dumped at sea off the coast of Australia.[20]

Given the large numbers of chemical munitions produced during the period 1915 to 1946 and the extent to which they were shipped around the world, it is not surprising that, despite the various disposal operations undertaken after both World Wars, these munitions continue to be found at regular intervals in a number of countries. Unlike the large chemical weapon stockpiles of the Russian Federation and the United States, which consist of a limited number of types of chemical munitions stored at a small number of locations, these old chemical munitions generally turn up in small numbers, in poor condition and at scattered sites. Their safe and efficient disposal therefore poses a very different problem to that of the US and Russian stockpiles. Before discussing some of these problems it perhaps first worth examining, briefly, the implications of the CWC with respect to these old and/or abandoned chemical weapons.

TABLE 5. United States chemical weapon depots and stocks in Australia during 1943

Location	Chemical Weapon Inventory
Charters Towers, Queensland	115 tonnes of bulk agents; 5,900 filled 100 lb. bombs; 1,000 empty spray tanks; and 600 empty bombs
Kangaroo (north of Townsville, Queensland)	5,500 "mustard' filled 100 lbs. bombs; more than 20,000 toxic artillery shells
Darra, Queensland	435 tonnes of bulk agents; 90,000 artillery rounds; and a small supply of toxic smoke candles, empty bombs, spray tanks and chemical land mines
Columboola (200 miles west of Brisbane, Queensland)	11,000 'mustard' filled 100 lb. bombs
Kingswood (near Sydney, New South Wales)	approximately 53,000 artillery shells (chemical)
Geelong, Victoria	400 tonnes of bulk 'mustard' and 3,160 toxic smoke candles

5. CWC Requirements

The conclusion of negotiations on the Convention on the Prohibition of the Development, Production, Stockpiling and Use of Chemical Weapons and Their Destruction (CWC) in the Conference on Disarmament in Geneva in August 1992 brought to an end 20 years of negotiations on a comprehensive treaty to eliminate chemical weapons. To date 160 countries have signed the Convention since it was opened for signature in Paris in January 1993, and 48 have ratified it. It will enter into force 180 days after the deposit of the sixty-fifth instrument of ratification.

States Parties to the Convention undertake a number of obligations with respect to any chemical weapons that they possess or that are present on their territory. Whilst the Convention requires that all chemical weapons shall be destroyed, it makes some distinctions between modern chemical weapons that continue to pose a significant military threat and those old and/or abandoned chemical weapons which do not. Chemical weapons produced after 1946 are considered to pose the highest threat to the object and purpose of the CWC, and as such are classed as chemical weapons. They are subject to declaration, and systematic verification of their storage and destruction by international inspectors from the Organisation for the Prohibition of Chemical Weapons (OPCW). Chemical weapons produced during the period 1925 to 1946 which have deteriorated to the point where they are no longer usable are classed as old chemical weapons. These weapons must also be declared and destroyed, but will generally be subjected to a less stringent verification regime by the OPCW. Chemical weapons produced during the period 1925 to 1946 and which are determined to be usable are classed as chemical weapons and must be treated in the same manner as the post 1946 weapons. Chemical weapons produced prior to 1925 are also classed as old chemical weapons. In this case, however, whilst they must be declared and destroyed, their destruction will not be subject to verification by the OPCW's inspectors.

The Convention requires that a State Party should destroy any chemical weapons it possesses within the first ten years after the Convention enters into force. In exceptional circumstances this period may be extended to 15 years. It is important to note that chemical weapons possessed by a State Party and buried on its own territory prior to 1 January 1977 do not have to be declared or destroyed as long as they remain buried. If, however, these weapons are recovered they must be declared and destroyed in accordance with the provisions of the Convention. Chemical weapons dumped at sea prior to 1 January 1985 do not have to be declared and are not subject to verification.

Special rules apply to chemical weapons that were produced by one State Party and abandoned on the territory of another State Party. Both States Parties are required to declare details of such weapons to the extent possible. Where the abandoning State Party is known or can be subsequently identified, it is responsible for the recovery and destruction of the weapons. In situations where it is not possible to identify the abandoning State Party, the responsibility for the recovery and destruction of the weapons falls on the State Party on whose territory the weapons have been abandoned. In this situation, however, the State Party may seek assistance from the OPCW.

Given the large numbers of chemical munitions produced during the period 1915 to 1946 and the extent to which they were shipped around the world, it is not surprising that, despite the various disposal operations undertaken after both World Wars, these munitions continue to be found at regular intervals in a number of countries. Unlike the large chemical weapon stockpiles of the Russian Federation and the United States, which consist of a limited number of types of chemical munitions stored at a small number of locations, these old chemical munitions generally turn up in small numbers, in poor condition and at scattered sites. Their safe and efficient disposal therefore poses a very different problem to that of the US and Russian stockpiles. Before discussing some of these problems it perhaps first worth examining, briefly, the implications of the CWC with respect to these old and/or abandoned chemical weapons.

The CWC requires that a State Party shall destroy its chemical weapons by an essentially irreversible process and that the products from such a process shall be unsuitable for use in the production of chemical weapons. The selection of the specific destruction method is left to each State Party, subject to the proviso that disposal by dumping in a body of water, land burial or open pit burning may not be used. The destruction facility using the chosen process must be constructed and operated in such a manner as to permit the destruction process to be verified in accordance with the provisions of the CWC.

6. Recovery and Identification of Old and Abandoned Chemical Weapons

Many countries routinely recover munitions from old battlefields, old test ranges and past storage sites. Belgium, France, Germany, Italy, the United Kingdom and the United States of America have well established programmes for the recovery and safe disposal of these weapons. A description of some of these programmes is provided in a number of the following papers, and others are reported elsewhere.[9][23][24]

One problem which is common to all finds of old chemical weapons is that, more often than not, their condition is such that it is not possible to tell immediately whether they are chemical or high explosive munitions. In situations where it is considered possible that the recovered munition might contain chemicals, the first task is therefore one of identification. It is sometimes possible, with very experienced explosive ordnance disposal (EOD) teams, to identify characteristics which make it possible to distinguish between chemical and high explosive rounds. Slight dimensional differences, the presence of a filling plug, or the particular munition or fuse design can all be helpful indicators. It is rarely possible, however, even for experienced EOD teams, to make an absolute determination on the basis of physical characteristics.

A well established tool for assisting with this identification is the use of X-ray techniques. Munitions may either be X-rayed at the site of the find, using portable equipment, or after removal to a disposal facility and after cleaning to remove surface debris and corrosion has been carried out. The X-raying of a munition enables vital information on its internal configuration and status to be obtained. Apart from providing information on the internal design and whether a liquid fill is present, the radiograph can also provide information on the explosive configuration and - where a munition is also fused - on its status.

In recent years some research has been undertaken on the development of other non-intrusive techniques for determining the contents and status of munitions. These include neutron activation analysis, ultrasonics and acoustic resonance techniques. Whilst primarily aimed at the Russian and U.S stockpiles, these techniques may nevertheless also have application in the old chemical weapons field. One technique in particular has been fielded and successfully used by U.S. EOD teams on a number of occasions over the last two years. This device, which is based on the use of photon-induced neutron spectroscopy (PINS), makes it possible to identify the presence or absence of the elements phosphorus, sulphur, chlorine, fluorine and arsenic within the munition. Clearly this information not only helps to identify the presence of a chemical fill, but in many circumstances also means it is possible to identify the fill. The development of more and improved techniques of non-destructive analysis will be a major factor in the development of safer techniques for the recovery and disposal of these old munitions.

One example of the problem of the recovery and identification of old WWI chemical weapons is provided by the Belgians. [24] More than 75 years after the end of World War I, explosive ordnance disposal units in Belgium still collect about 3000-3500 munitions per year from the Flanders region. The total weight of munitions collected each year averages 200-300 tonnes of which about 10%, or 20 tonnes per year, are subsequently confirmed to be chemical munitions. A complete range of ammunition belonging to all the nationalities which participated in WWI is found. An added complication is that the munitions have been fired, and are therefore not only fuzed but potentially still armed.

After recovery, the munitions are taken to the Belgian storage and disposal facility at Poelkapelle, in Flanders, where they are first cleaned to remove debris and the products of corrosion. A tentative identification is then made with the assistance of a diagnostic chart developed by the facility itself, as well as of a specially constructed X-ray facility. Until 1980 cleaning was done by carefully tapping the side of the munition (not the fuze!) with a hammer and chisel. Later this was first replaced by a manual system using a high pressure water jet, and then by a high pressure sand-water mixture. In a new cleaning/identification facility currently under construction a jet of frozen carbon dioxide pellets will be used to clean the munitions.

In this new facility cleaning and identification will be undertaken by remote means. The new building is basically divided into two halves separated by a blast wall. One half houses the control room and support area, while the other half houses the munition handling area. Munitions will be loaded manually onto specially designed pallets at the entrance to the cleaning area. The pallets will then be moved under remote control to the cleaning station, where the munition will be cleaned using a blasting gun operating on solid carbon dioxide pellets. The cleaned munition will be moved from here to the radiography cell, where details of its internal structure will be determined and recorded. Finally the munition will be moved to a marking area where it will be given a unique identification code. All these operations can be managed remotely from the control room. The operating area is also being designed to cope with leaking munitions or, in the worst case, a detonation. After identification the high explosive munitions will be destroyed in the normal manner and the chemical munitions stored until they can be disassembled in the site's dismantling facility.

Once a munition has been identified as a chemical munition, and where it is explosively configured, the status of the explosive components determined the most appropriate method for its safe disposal can be determined.

7. Disposal of Old Chemical Munitions

Chemical munitions without an explosive or propellant component pose a lesser hazard and can generally be treated in a similar manner to storage containers of bulk agent. On the other hand, the situation with CW munitions which contain explosives or propellants is entirely different. Destruction of these items in a safe and environmentally acceptable manner can pose major difficulties. The historical solution to this problem has been to deep sea dump. As this option is no longer available, there are only two basic approaches which can be used: (a) the explosive component must be separated from the CW component and then each element must be destroyed by an appropriate safe method; or (b) processes capable of safely handling the combined hazard must be developed. The approach being used by all countries currently undertaking disposal is option (a). The problem is that the disassembly of old chemical munitions, which are generally in a very poor condition, is both very hazardous and time consuming. The use of remote handling techniques, wherever practicable, can help to reduce the hazard, but still normally results in the munitions having to be handled one by one. A typical daily throughput for such a disassembly operation will range from 15 to 30 munitions.

One alternative approach to the problem which follows option (b) is the cryogenic fracture process currently under consideration in the United States for possible use in their chemical demilitarisation programme. In this process the complete munition is cooled in liquid nitrogen, to make it brittle, and is then mechanically fractured into small pieces. These pieces can in turn be fed directly into a high integrity incineration plant. The system has been designed to provide complete chemical and explosive containment. Plants of this type, however, require sophisticated technology and engineering and the safety of this technique, when used with explosively configured old chemical munitions, has yet to be established fully. If it can be shown that this technique is effective, it should be capable of providing a relatively high throughput with a good degree of safety and environmental protection.

Better techniques for preparing old chemical munitions for disposal are clearly needed. These techniques must not only cope with the safety, environmental and throughput problems, but must also need to address the need for portable or transportable systems suitable for using in the field.

Once the problem of disassembly has been solved, the next step is to dispose of the chemical agent and, where necessary, of the explosive component. Over the years considerable experience has been gained with the disposal of "mustard gas" by high temperature incineration, and this method has been successfully used by Canada, Germany, the United Kingdom and the United States. The major advantage of incineration is that it can readily deal with "mustard gas" which has been thickened or become polymerised over time. It can also handle decontamination of the associated metal munition components. Some experience with alternative techniques for the disposal of "mustard gas" have been obtained. The Russian Federation has successfully used a process involving the hydrolysis of "mustard gas" using monoethanolamine. [25]. Final disposal of the hydrolysis product was still carried out by incineration.

The disposal of arsenical based chemical agents has proved to be more problematical. Here the principal difficulty is that the final product will still contain arsenic in some form or other and will therefore still be toxic, albeit to a lesser extent. The Italians have developed and successfully used a technology for disposing of bulk stocks of a mixture of "mustard gas" (50%), phenyldichloroarsine (45%) and arsenic trichloride (5%). The material was treated with hydrogen peroxide under alkaline conditions and the product immobilised in concrete. The facility, which is located at Civitavecchia, near Rome, has a destruction rate of approximately 250 kg per batch. [23] The Canadians, as part of their Project "Swiftsure", destroyed 1.5 tonne of Lewisite by oxidation followed by neutralisation. In this process the Lewisite was neutralised using an alkaline peroxide solution in the presence of sodium or calcium chloride. The reaction yielded solid sodium or calcium arsenate, gaseous acetylene, and dissolved chloride salts, and these products were again immobilised in concrete. The plant was skid mounted and could accept the agent directly from a one tonne storage container.[26] Germany is building a new incinerator plant specifically designed to process chemical agents containing arsenic. Details of this are provided in one of the following papers. Notwithstanding these three successful operations, the disposal of arsenical containing chemical agents would benefit from further research.

Whichever disposal method or methods are finally chosen, the aim must be to achieve appropriate high standards of safety and environmental acceptability. It must be remembered, however, that these two requirements cannot be considered in isolation from each other. The safest and most expedient disposal method may not necessarily be the most environmentally friendly. Whilst there is a clear need to protect the environment, the safety of the personnel directly involved in the operation must be paramount. In difficult and hazardous operations such as these it is essential that external parameters, such as environmental emission standards, are set at realistic and scientifically justifiable levels. Apart from making the operation both more expensive and protracted, unjustifiably high standards can actually increase the immediate risk to the personnel undertaking the work. Some form of compromise will therefore inevitably be necessary.

8. Summary

In this brief overview the author has tried to explain how the problem of old and abandoned chemical munitions has arisen, to give some idea of the extent of the problem and to indicate what some countries are doing about the problem. The author has also tried to highlight those areas where further research is required such as:

(a) new or improved techniques for the non-destructive analysis of the internal status and contents of old munitions;

(b) alternative methods for the preparation of old chemical munitions for safe disposal;

(c) the development of destruction methods that can cope effectively with thickened or polymerised "mustard gas" ;

(d) the development of environmentally acceptable destruction processes for chemical agents containing arsenic; and

(e) the provision of additional data on the environmental impact of chemical agents and their degradation products.

9. References

1. Hemsley, J. (1987) *The Soviet biochemical threat to NATO.* St Martins Press, New York.
2. Hanslian, R., et al. (1937) *Der Chemische Krieg*. 3rd ed., Berlin
3. Thomas, A. (1985) *Effects of chemical warfare:* a selective review and bibliography of British State Papers. SIPRI, Stockholm
4. Prentiss, A.M. (1937) *Chemicals in War.* New York.
5. Zanders, J.P. (1993) *The destruction of old chemical munitions in Belgium,* in T. Stock & K. Lohs (eds.), *The challenge of old and abandoned chemical weapons,* SIPRI, Stockholm (in press).
6. Kleber, B.E. & Birdsell, D. (1990) *The chemical warfare service: chemicals in combat.* Center of Military History, United States Army, Washington DC.
7. Peoples Republic of China Working Paper (1992) Some information on discovered chemical weapons abandoned in China by a foreign State. CD 1127, Conference on Disarmament, Geneva
8. Indonesia and the Netherlands Working Paper (1982) Destruction of about 45 tonnes of 'mustard' agent at Batujajar, West Java, Indonesia. CD 270, Conference on Disarmament, Geneva.
9. Manley, R.G. (1990) The safe disposal of hazardous military chemicals. Chemical waste incineration, 12-13 March 1991, Manchester.
10 Crone, H.,Wailes, P., Douglas, J. & Abbot, N. (1991) Johnston Atoll chemical weapons disposal programme: South Pacific Forum scientific evaluation. Materials Research Laboratory, Victoria, Australia.
11. Garrett, B.C. (1993) Abandoned CW stocks uncovered in Washington, DC. *ASA Newsletter,* issue 35, 8 April 1993.
12. SIPRI, (1971) *The problem of chemical and biological warfare: the rise of CB weapons*, vol. **1**, Stockholm.
13. Brophy, L.P., Miles, W.D. & Cochrane, R.C. (1959) *The chemical warfare service: from laboratory to field.* Office of the Chief of Military History, United States Army, Washington DC.
14. United Kingdom Working Paper (1988) Past production of chemical warfare agents in the United Kingdom. CD 856, Conference on Disarmament, Geneva.
15. Pearson, G (1995) Farewell to arms, *Chemistry in Britain* **31** 782-786.
16. Ad Hoc Working Group on Dumped Chemical Munition (1994) Report on chemical munitions dumped in the Baltic Sea to the 15th meeting of the Helsinki Commission, 8-11 March 1994.
17. Federov, L. (1995) Undeclared chemical weapons in Russia. Centre for Russian Ecological Policy, Moscow.
18. CW Plant behind schedule: Environmental Threat, JPRS Proliferation issues (original Izvestiya), JPRS-TND-94-006, 16 March 1994
19. SIPRI, (1994) *Yearbook,* Oxford University Press, Oxford.
20. Haug, M. (1995) A preliminary survey of open source information on old and abandoned chemical weapons of some countries in the Asia/Pacific region, (in press).
21. Report on sea dumping of chemical munitions by the United Kingdom in the Skagerrak waters post World War II, CHEMU 2/2/5, submitted by the UK to the Ad Hoc Working Group on Dumped Chemical Munitions, Helsinki Commission, 28-30 September 1993.
22. Wiseman, D.J.C., (1951) *Gas Warfare.* vol. **1**, The War Office, London
23. Italian Working Paper (1991) Italian experience of the destruction of old and obsolete chemical weapons. *CD/CW/WP.* **375**, Conference on Disarmament, Geneva.
24. Wouters, I.R. (1991) Old and obsolete chemical weapons: technical problems of dismantling. Presented at the meeting on the destruction of chemical weapons 8-10 October 1991. Conference on Disarmament, Geneva..
25. Union of Soviet Socialist Republics Working Paper (1991) Main technological aspects of the destruction of chemical weapons. CD/CW/WP. **367**, Conference on Disarmament, Geneva.
26. McAndless, J.M., Fedor, V. & Kinderwater, T (1992) Destruction and waste treatment methods used in a chemical agent disposal project. Suffield Memorandum No. **1392**, Defense Research Establishment Suffield, Canada.

DESTRUCTION OF OLD CHEMICAL MUNITIONS

DANIEL FROMENT
Departement des Recherches, Etudes et Techniques
Délégation Générale pour l'Armement
91710 Vert-le-Petit, FRANCE

1. Treaty Obligations

The problems posed by the elimination of old chemical deposits that still exist within a country's territory can no longer be addressed at a purely national level, but must be considered in the light of the liabilities ensuing from the signature and ratification by the member state of the text of the Convention on the Prohibition of the Development, Production, Stockpiling and Use of Chemical Weapons and on their Destruction.

This treaty, signed in January 1993 in Paris, contained two clauses referring to the elimination of chemical arsenals and the destruction of old or abandoned chemical weapons. According to the provisions of the Convention, a member state that uncovers old chemical weapons after the convention has come into force must declare them to the technical secretariat within 180 days of their discovery. The technical secretariat is fully empowered to verify the accuracy of the information supplied. It goes without saying that the elimination of this old waste is a moral obligation. This must be achieved using appropriate facilities, and respecting safety and environmental considerations; in principle it should be possible to get help from the originating country if the munitions had been abandoned.

2. The Problem of Old Munitions

2.1. PRODUCTS USED

The combatants in the 1914-18 war did not pre-plan the use of toxic gases on the battlefield. It was only after the front stabilised at trench lines that the requirement arose to make these positions untenable. Recourse to chemicals appeared to be the best means of achieving this aim. In a very short time more than 3000 compounds were therefore identified and examined to see if they could be used as war toxics, with most showing little or no effectiveness. Only about 30 were assessed as suitable for field use, and 12 were adopted and used intensively. Out of these 12, 6 finally gave satisfactory results.

J.F. Bunnett and M. Mikołajczyk (eds.), Arsenic and Old Mustard: Chemical Problems in the Destruction of Old Arsenical and 'Mustard' Munitions, 17–32.

It is interesting to note that, between the two wars, 1400 products were tested in France and 2 were finally selected. The total number of substances examined by the world's chemists for possible use as attack agents, since 1915, which is 81 years ago, has been estimated at 12,000.

The reason for this meagre outcome, in relation to the effort expended, is to be found in the numerous requirements that a chemical product has to meet before being considered as a war toxic with a reasonable chance of successful use. Putting aside technical matters such as supply difficulties, stability and cost, it was necessary to consider tactical considerations such as persistence, insidiosity, volatility, etc. For the sake of convenience and because of lack of time, most of the agents chosen came from amongst compounds that were basic to the chemical industry.

The table below lists the main agents used by the various combatants (France, GB, Germany) during the first World War, as well as a rough figure, probably underestimated, for the quantities used.

AGENTS	QUANTITIES (tonnes)
Chlorine	65000
Phosgene	30000
Vincennite	4160
Diphosgene	13000
Yperite ('mustard gas')	11000
Clark I	4200
Chloropicrin	8000
Bromoacetone	481
Acrolein	183
$AsCl_3$	2710

To these products must be added 4116 tonnes of tin chloride and 191 tonnes of titanium chloride used as smoke generating agents by France. Great Britain used virtually the same agents as France.

2.2. EVALUATION OF NUMBER OF ITEMS

The number of loaded shells per combatant country breaks down as follows:

FRANCE	16 MILLION
GERMANY	33 MILLION
GREAT BRITAIN	4 MILLION
USA	1 MILLION

Shells of all calibres (75 to 210 mm) were the main type of munition used. Others were mortars and bombs of various sizes, with the explosive charge varying from about a hundred grams up to 8 kg.

A number of shells were transferred, mainly from France to allies:

BELGIUM	193000
USA	948000
GREECE	12000
ITALY	92000
PORTUGAL	46000
RUSSIA	12000
ROUMANIA	50000

The percentage of misfires was usually between 20 and 25%, which means that at a rough estimate more than 10 million objects may have been left buried on old battlefield sites. Nearly all these weapons are fused. The origin of these munitions is: German: 60%, French: 30%, and British: 10%. The inventory carried out shows that more are than 100 different types of munition are involved.

2.3. THROUGHPUT OF THE INSTALLATION

It is important to have an estimate of the quantity and type of objects to be processed. To a large extent this quantity is dependant on economic activity inasmuch as major construction projects (roads, motorways, bridges, railway lines, buildings) result in most of the discoveries. On average, one kilometre of railway line or motorway in a "sensitive" area generates ten tonnes of "waste". The location and number of "unofficial" dumps is usually unknown, as most of the archives were destroyed, and witnesses are becoming increasingly rare. The proportion of munitions described as "chemical" is assessed to be less than 10%.

Only one compound ("mustard gas") appears in Schedule 1 of the Convention. The other agents, after removal from their container may, in principle, be destroyed by treatment in an approved industrial chemical waste plant. It is clear that recovery of these compounds is a difficult or even impossible task, and offers no benefit.

Design of a plant for the destruction of old and/or abandoned chemical munitions presents problems at several levels.

The quantity of objects requiring "neutralisation" is unknown, and hence there is a problem of establishing a throughput for the unit. Treatment rates in current plants vary from a few shells to about twenty per day. These projectiles, usually in very poor condition, present both chemical and explosive hazards. The supply is "random" and can only decrease with time. Will the plant operate continuously or only in batches? Will it destroy the agent and the casing simultaneously? When compared with the weight of products usually handled by the chemical industry, the quantities requiring destruction are very low (a few tonnes per year).

2.4. THE PROBLEM OF ARSENIC

The presence of arsenic adds significantly to the complexity of treatment for these compounds. The compounds of German origin, Clark I and Clark II, were sometimes mixed with explosive (Clark II was used in the very final months of the war). France used a lot of arsenious chloride mixed with prussic acid (HCN). Arsenic cannot be discharged into the soil and must be subjected to special treatment. The estimates given above suggest a weight per year of 500 kg. The most widely used process involves precipitating the arsenic in the form of a salt and storing it in old mines.

2.5. PERSONNEL TRAINING

This must be taken into consideration from the very beginning of the project. Destruction of chemical weapons is a very specialised activity. Few people are acquainted with these objects, which usually induce a powerful feeling of distaste. The training of personnel required to work at a site must be considered at an early stage. American specialists have built a full-scale replica of the installation to be used to dismantle their arsenal. More than two thousand personnel will be trained in the various techniques used in the destruction plants that will have to operate on American soil. Training involves 18 lectures covering mainly electrical circuitry, analytical instrumentation, sampling, incineration, the munitions, laboratory equipment, maintenance, protective equipment, and safety, monitoring and detection systems.

3. Environment and Communication

Current legislation in the countries concerned may impose additional and not inconsiderable constraints on the choice of a process and the location of destruction installations. Account must be taken not only of standards governing emission limits and rules applicable to the transport of munitions, but also of any pressure that environmental protection movements might apply.

Any country or organisation involved in chemical weapon destruction activity must ensure that it has access to the information needed to appreciate the relation between problems of disarmament and those of the environment. There are tools that can be used to take account of this environmental dimension, which will then be incorporated into the destruction procedures adopted.

3.1. THE IMPORTANCE OF PUBLIC OPINION

The value systems in most contemporary cultures and civilisations are unwilling to accept exposure of citizens to pernicious and insidious poisoning. It is essential that the elimination of these weapons be carried out in an irreproachable fashion to ensure that the organisations responsible for this task cannot be accused of contaminating the population or the environment. It is for this reason that institutions responsible for the destruction and for its monitoring must apply objective criteria to enable a quantitative assessment of the impact on the outside world of the processes employed. (Such an assessment might also de-fuse hostility that might be generated by such operations.) Such discussions are complicated by the sometimes unavoidable confidentiality associated with agent destruction.

3.2. ENVIRONMENTAL MANAGEMENT TOOLS

Two types of tool are used to identify and manage the environmental impact of industrial activity in the broadest sense of the term. These instruments are the "écobilan," which registers the flow of material and energy associated with an industrial process, and the site audit, which identifies environmental problems associated with the industrial site.

3.2.1. *The Ecobalance ("écobilan")*

This is a statement of material and energy produced at the boundaries of a given system. The "écobilan" usually forms part of an investigation entitled "life cycle analysis", which itself breaks down into four stages:

- definition of targets. The problems of the sponsor are translated into terms that will enable the "écobilan" to provide quantitative answers
- evaluation during the life cycle. This is the ratio of energy/material consumed or discharged throughout a product's life cycle
- environmental impact. Translation of this physical ratio into an environmental impact
- possible improvements. Follows from the results of the preceding two steps.

For the destruction of chemical weapons, the "écobilan" must be supplemented by a toxicological approach and a risk analysis.

3.2.2. *Site Audit*

The site audit covers the environmental status of a site and examines the contributions made to any environmental change.

A first phase involves assembling and analysing available information, discussions with managers and a site visit. The following important points must be taken into account:

- site activity liable to create pollution or environmental damage
- a hydro-geological survey of the site
- operating restrictions resulting from clearances and regulations
- the flow rate of any substance that might be the cause of any pollution
- a check of site monitoring data.

A second phase involves reviewing conformity of operations with clearances and regulations and with internal company standards.

Finally, a summary of the impact of the industrial activity is produced, and costed proposals for any site improvement or rehabilitation are provided.

3.3. CHEMICAL WEAPONS DESTRUCTION AND THE ENVIRONMENT

The organisation with responsibility for the destruction of chemical weapons must have the facilities required to carry out environmental monitoring of the destruction processes and systems.

The "écobilan" is particularly well adapted to the monitoring of any discharge caused by an industrial activity, throughout the life of the process.

The site audit enables the environmental status of a site being used for the destruction of chemical weapons to be monitored. It can also be used once the destruction has been successfully completed, to help restore the site for other uses, and, when appropriate, to enable solutions for its restoration to be recommended.

The environmental dimension of this destruction activity must be taken into account to achieve maximum control of the conduct of the elimination operations and of their potential environmental impact, to increase protection for both personnel and for the ecosystem and to present an irreproachable media image.

4. Task Organigram

Resolution of this complex problem begins with an organigram (a task organization chart) for the tasks to be carried out from the moment the munition is uncovered until its final elimination.

4.1. TASKS IN THE FIELD

4.1.1 *Detection and Discovery*

Discoveries are usually of a random nature (farm worker in his field, earth moving operations) unless archival information enables some direction to be given to any investigation. Searches carried out in the area of old production or storage sites are often profitable. There is some mapping of munition type by combat zone which enables fairly accurate targeting by content type.

4.1.2. *Cleaning*

To obtain as much information as possible about a buried object, one should clean it while checking for leaks with a detection apparatus. It is often possible to find identification marks visually or to recognise the type of weapon (presence of a filling plug for example). Cleaning using a water jet at high pressure seems to be the method to use. An alternative is the use of solid CO_2, in sand-like condition; it is put into the drained casing, pushed around with a tool, and dumped out. It is quite effective.4.1.3.

4.1.3. *Identification*

If there is doubt as to the type of munition recovered, the use of some form of non-destructive assessment technique is essential. X-ray examination usually enables distinction between a conventional and a chemical round, and tomography provides accurate information on the internal dimensions of an object if penetration or sectioning is being considered.

Neutron activation analysis can indicate the nature of the constituent elements. It enables the creation of a database and the sorting of projectiles as a function of elements contained (the presence or absence of arsenic, for example). It is now possible, using this technique, to identify chlorine or arsenic in less than two minutes. The neutrons may come from a natural source (such as Am-Be or Cf) or from a generator tube. In both cases, a mobile analyser is desirable.

4.1.4. *Transport*

It is essential that transport conditions offering a high degree of security are specified. The very poor condition of most of the munitions means that every precaution must be taken to avoid accidents when the objects are in transit. Development of refrigerated containers, associated with a vehicle with reinforced structure may be the answer to this problem. Recourse to refrigeration is all the more justified as a large proportion of the agents used contain highly volatile raw materials such as chlorine (b.p.= -34°C) or phosgene (b.p.= 8°C). In Livens bombs, which contained 12 kg of phosgene, internal pressure can reach 6 to 7 atmospheres.

When the munition is assessed as being un-transportable, it must be destroyed on site in accordance with existing or yet-to-be–defined protocols.

4.2 TASKS AT A CENTRAL FACILITY

4.2.1. *Unloading*

Unloading at the storage site would not appear to require any special facilities, apart from a respect of procedures and the need to provide equipment for decontamination of the containers and the transport vehicle.

4.2.2. *Identification*

Definitive information must be obtained about a munition before it can be dispatched to its storage location. Neutron activation analysis and tomography equipment must be available on site, for purposes as discussed above. In some cases, samples of liquid munition contents are taken by drilling into it with a hollow drill. It is necessary that conventional chemical laboratory analysis facilities (mass spectrometer, chromatography, data base, etc.) be available so that any suspect material can be identified. The chosen dismantling techniques will be all the more effective if they are applied to well defined families of objects.

The creation of a data base with information on all munitions used as well as their characteristics is nearly completed.

4.2.3. *Storage*

Before destruction, the munitions are stored in concrete hangars, which are sealed and equipped with filtration systems to counter both chemical and explosive hazards. Fixed detection equipment must be installed with a centralised remote alarm system. Definition of storage capacity per cell is difficult as it is practically impossible to study the behaviour of this type of munition (drop test, wave propagation, state of the fuse, etc.). Nevertheless, it is possible to evaluate munition sensitivity on the basis of the following criteria:

- munition characteristics
- list of possible rupture modes
- analysis of the incidence of fuse triggering following normal and accidental inputs
- analysis of sympathetic detonation
- analysis of fuse triggering in the case of fire
- analysis of non-pyrotechnic rupture modes

Prevention of any propagation of incidents is a major criterion in the design of munition storage methods.

The various techniques that can be used for dismantling and destruction are summarised below without prejudice to any future application.

5. Dismantling

There are many techniques available for the dismantling of chemical weapons. Some are already in use at existing installations, others are at the exploratory stage and will probably be developed later.

Dismantling is the combination of operations (simultaneous or sequential) that results in the physical separation of the fuse, the explosive charge, and the toxic agent. The operations that form part of a process may be implemented simultaneously or stepwise. It is therefore clear that, if a wide range of munitions is to be processed, several dismantling workshops are required.

5.1. THE MECHANICAL OPTION

Removal of the fuse from the pyrotechnic sequence may be achieved by power-assisted unscrewing. As most of the objects are in poor condition, this is not a realistic option.

5.2. DRILLING AND ASPIRATION

The casing is penetrated using a drill bit, and the contents are then sucked out for processing. This technique is applicable to all munitions containing a liquid and whose internal geometry is known. It is possible to design a tool including an expansion chamber to contain any compressed gas from within the projectile, which enables all munitions containing compressed gases such as phosgene to be tackled. The operation may be completely automated. It is used by the USA, Germany, the UK and Italy.

This process cannot be used on munitions containing solid agents.

5.3. SECTIONING

The munition is sliced in two to give access to its internal components. This can be applied to all types of munitions, and is essential for those with a solid charge.

Treatment rates are low (a few units per day). A fluid is required to cool the munition during cutting. Internal geometry of the munition must be fully known. The risk of munition initiation is not negligible.

5.4. CHEMICAL DISSOLUTION

The munition is dissolved by use of a corrosive agent.

This technique may enable the simultaneous disassociation of the energetic components, decontamination of the casing and destruction of the toxic agent; it can be applied to all situations that do not lend themselves to conventional opening methods. It may be used for 210 mm projectiles weighing 120 kg and containing up to 8 kg of explosive as well as for the disassociation of the energetic components extracted after sectioning by means of a water jet.

The time taken for chemical dissolution and the behaviour of the explosive in a corrosive bath must be known before progressing to an industrial application, as well as the behaviour of the bath in the presence of the primary explosive picrate. It will be necessary to process solutions containing high concentrations of mineral salts and metal waste. This minimally aggressive technique has the inconvenience of a low throughput. Good results are obtained with the following mixture: $FeCl_3$ (950 g); Fe $(NO_3)_3 \cdot 9H_2O$ (950 g); HNO_3 (965 ml); H_3PO_4 (1195 ml); H_2O (2840 ml). One litre of solution dissolves about 60 g of metal.

5.5. ELECTROCHEMICAL DISSOLUTION

A variant of chemical etching, electrochemical dissolution based on the principle of the dissolving anode, being a carefully selected part of the munition, offers the theoretical advantage of a wide choice of electrolytes (acids, neutrals or bases) whose composition may neutralise or even destroy the explosives and/or chemical agents.

This technique enables the casing to be penetrated in complete safety and it can be adapted to all situations that do not lend themselves to the use of conventional opening techniques. The electrode shape can be adapted to produce any shape of cut-out. This process seems to have progressed beyond the laboratory stage as a UK company (AEA) has proposed it at specialised conferences.

5.6. SECTIONING USING AN ABRASIVE JET

The casing is cut using an abrasive jet (water or decontaminant) in order to extract the energetic components. This process is suitable for 155 mm shells with a large explosive charge (greater than 100 g) and solid or liquid contents.

This technique enables damage caused following an explosion to be limited as the cut is carried out on a munition immersed in a bath that can neutralise the chemical agent immediately. It is possible to recover the pyrotechnic chain automatically. A shell can be sectioned in 3 minutes using 10 litres of fluid. This technique is experiencing significant growth.

This equipment is very noisy in the open air. A mixture of water/explosive/toxic/abrasive may be generated. It would seem sensible to combine its use with vacuum drilling extraction in order to treat a munition that has been partially emptied of its contents.

5.7. PENETRATION USING A HOLLOW EXPLOSIVE CHARGE

The munition is penetrated using a mini hollow charge.

This process, which is applicable to any munition containing an un-pressurised non-inflammable liquid, with internal geometry fully understood, enables rapid opening to be achieved. Applied to a series of well known and identified munitions and associated with aspiration, it results in the generation of a slightly contaminated casing whose explosive charge is then isolated by cutting using a water jet.

Application of this technique to a munition with fuse chain already primed can cause explosion of the munition. If there is high pressure within the munition, a "champagne" effect in the contents may be provoked; there is also a risk of fire.

5.8. CRYOFRACTURE

After immersion in liquid nitrogen, the munition casing becomes very brittle and liable to rupture or fragmentation when force is applied using a hydraulic press.

This process can be applied to most munition types. Exceptions are munitions with aluminium casings.

Cryofracture seems to be a very promising technique as it creates a degree of "insensitisation" of the weapon. The solidification of the toxic agent reduces the vapour risk considerably.

Currently there are no results available concerning the processing of old munitions using this method. However this technique is costly and requires significant numbers of objects for destruction in order to be cost effective. It must be combined with incineration. The presence of frozen "lumps" may lead to uneven combustion and erratic furnace operation.

5.9. OTHER TECHNIQUES

For many years the prevailing geopolitical situation has favoured disarmament activities. Industry is also interested in this new field and now offers some very original techniques and products. These include explosion in a vacuum chamber weighing 120 tonnes and with a volume of several tens of cubic meters, the plasma torch, cold sectioning etc.

6. Destruction

Although the choice of destruction technique is critical and fundamental, a certain number of considerations may influence decisions: only one compound ('mustard gas') appears in Schedule 1 of the Convention. If yperite is drained and collected in a drum, it becomes a modern weapon and the facility may be subject to annual inspection. It must be destroyed using a specialised plant subject to monitoring. The other agents (which must be identified) may be destroyed using an approved industrial waste processing plant. However, the product still has to be extracted from the munition.

6.1. CHEMICAL DESTRUCTION

Chemical destruction is usually achieved using hydrolysis or oxidation. It is the first process to have been used. This technique has the disadvantage of not being very effective on a large scale as the reaction time may be fairly long if an acceptable extent of conversion is to be achieved (> 99%). The volume of waste generated is significant (a ratio of 1 to 6), waste that then has to be concentrated or incinerated in a dedicated plant. Finally, decontamination of the containers (which is a rapid operation using incineration) is very labour intensive, and may even be impossible. The problem of crusty mustard is not yet solved.

6.2. INCINERATION

This process involves burning the toxic at between 1000 and 1200°C in a flow of oxygen. The incineration results in virtually complete destruction of the products to be eliminated. It is a well known technology, generally proven and tested, covering a wide range of chemical families, and for which the installations are neither excessively large nor subject to environmental constraints (climate, modest land requirements, confinement, safety etc.). The conversion rate exceeds 99.9999% (US information).

Products to be eliminated are converted into simple inorganic gases; packaging and containers are transformed into ash or metallic residue. Complete combustion requires careful control of toxic agent/oxidant/fuel proportions, contact time (from 1 to 3 seconds) and combustion temperature in order to minimise the quantity of by-products produced, some of which are undesirable. Gaseous mineral effluents are trapped in caustic scrubbing towers to create various mineral salts ($NaHSO_3$, $NaHCO_3$, Na_2SO_3, Na_2CO_3, NaCl, NaF, Na_3PO_4, Na_3AsO_3 etc.). After drying, these salts are disposed of in the environment or in disused mines.

The temperatures reached and the corrosive action of the gases generated cause rapid and significant wear to certain incinerator components, which require frequent replacement. Maintenance of these installations has to be taken into account when choosing a process.

This process seems to be unavoidable in the case of old chemical weapons.

6.3. PLASMA TORCH

For about twenty years, for mostly military reasons, the plasma torch technique has been the subject of a certain amount of development. In addition there are civilian applications for this process, either in steel making where it enables greater operational flexibility and improved blast furnace operation, in the car industry for the production of cast iron for engine blocks, or in the processing of waste. The high temperatures reached enable the decomposition of all molecules, even the most stable, the elimination of unburnt residues and the vitrification of solid waste products thereby rendering them harmless to the environment. Pyralene type PCBs are destroyed without difficulty by the plasma torch. Similarly, a mobile processing unit for asbestos exists which enables elimination of this mineral, which is so toxic to man. Although no application to chemical warfare agents has yet been found, this very reliable technique must not be forgotten, with its media image generally acceptable to the public.

6.4. MILD PYROLYTIC DEGRADATION

The binary CW agent concept involves using the munition as a container for agent precursors and as the production reactor vessel. It was tempting to reverse this concept and consider the munition as a destruction reactor vessel. This technique is particularly well matched to certain munitions filled with modern organophosphorous toxins such as VX or GD without explosive charge, or to old munitions containing less than 100 g of explosive.

The fired but unexploded munition is placed in a chamber heated to between 300 and 500°C to provoke the initiation of its charge. This chamber is adjacent to an incinerator which is used to destroy the released agents.

Throughput could reasonably be expected to reach 200 munitions per day, or nearly 40,000 per year. This technique effects destruction of the explosive, decontamination of the casing, and release of the toxic agent in a single operation. Studies into this process are still at an early stage. The size of the explosive charge and the temperature and time needed to decontaminate the casing must be known with some certainty.

6.5 . OXIDATION IN SUPERCRITICAL WATER

When subjected to supercritical conditions (pressure exceeding 218 bars and temperature over 374°C), water loses its polar properties, its density becomes 0.3, and its viscosity becomes comparable with that of a gas. It does not dissolve mineral compounds, but dissolves organic molecules.

Organic materials in solution in water under supercritical conditions are oxidised by oxygen introduced into the reactive medium. This process is comparable with medium temperature incineration (400 to 500°C). The by-products obtained are simple mineral compounds (N_2, CO_2, H_2O, HCl, SO_2 etc.). The addition of soda enables precipitation of the acids in the form of salts as a result of the low solubility of salts in water in these conditions. Efficiency exceeds 99.99%.

6.6. WET AIR OXIDATION

Water possesses destructive properties even below critical conditions. A company has developed an elegant process for the processing of liquid residue from household waste. The sludge is injected into a 2000 m deep well. The temperature reached (180 to 200°C) and the pressure (200 bars) are sufficient to destroy all organic compounds.

This process has aroused much interest and has led CIBA-GEIGY to build two plants operating on the wet cycle oxidation principle (300°C, 200 bars). Unfortunately the operating conditions are such that only compounds containing C, H, O and a little chlorine can be processed without causing rapid corrosion of the equipment. The residues must be pumpable or be placed in aqueous solution; chlorine is only acceptable in very small quantities, and elements other than C, H, N, and P have to be avoided. To date, it appears that these tests have not been carried out with actual toxins. Although very attractive on paper, it seems unlikely that this process will ever see any application in the field of chemical toxin destruction.

6.7. MOLTEN SALT PROCESS

The organic product is destroyed by a bath of molten salt (sodium carbonate alone or a mixture with potassium carbonate). The bath is stable, cheap, with no vapour pressure, non toxic and recyclable. The reactor temperature is between 800 and 1000°C. Recycling enables any unreacted salt to be reused. After dissolving and filtering of the carbonate solution, the addition of CO_2 enables a re-precipitation of the sodium bicarbonate.

6.8. BIODEGRADATION

Various processes and techniques have been examined and developed to achieve the biodegradation of pesticides; activated sludge, drop filtration, lagoon oxidation, anaerobic digestion. A certain number of enzymes possess the ability to hydrolyze organophosphorous insecticides. Being specific, they have names that derive from the hydrolised compound: sarinase, somanase, tabunase.

A lot of work was dedicated to this subject once it was found that certain bacteria were capable of proliferating on pesticides. It appears that processes applicable to chemical warfare agents have not been the subject of any initial industrial development. This technique could only be developed for contaminated soil or for water and would appear to be difficult to apply to old munitions.

6.9. DESTRUCTION USING ELECTROCHEMICAL ACTION

It is well known that silver in the +2 oxidised state is one of the most active known oxidants. Trials carried out on conventional organic products, and then on toxics, have demonstrated the attractions of this reagent. A pilot unit using this principle has been developed by a UK company. It would appear that there are no applications to old munitions.

6.10. ADIABATIC DESTRUCTION

A mixture of diesel oil/toxic may be used as fuel for a diesel engine. Raised to a temperature of about 1000°C, this adiabatic destruction is virtually total (Russian information). This is a very crude but effective process for the destruction of small quantities without the use of major installations.

6.11. EXPLOSION-WATER TANK OPTION

This process involves generating the explosion of a chemical weapon within a decontaminant liquid. Trials carried out at CEB have proved the feasibility of this process for the destruction of yperite. A "water tank" of about 30 cubic metres, belonging to a company specialised in the elimination of industrial waste, has been operating on this principle for many years in France. However, it is only right to raise certain points concerning the limitations and inconveniences of this method when applied to chemical warfare agents:

- The productivity is low (assessed as 4 shells per hour)
- The problem of munitions whose explosive charge has been removed
- Inapplicability to treatment of agents thickened using resins or that are highly viscous
- Certain toxins contain arsenic within their molecule. This process would disperse this element in a very large volume which would greatly complicate the essential recovery of this element
- It would be necessary periodically to remove the contaminated metallic fragments, which would be an onerous task
- Decontamination of the casing would not appear to be possible
- The gaseous bubble generated by the explosion is large

Interest in this process is now minimal.

6.12. MISCELLANEOUS

Other technologies that have also been suggested include photochemistry, superheated steam, the use of fluidised beds, etc. It would appear that no industrial application aimed at chemical warfare agents has yet seen the light of day. The pressure of public opinion and the environmental lobby are the forces driving research into technologies other than those that already exist. With the complexity of the problem, the need to validate any process using actual agents, the time-scale imposed by any industrial development and the potential risks posed by any ageing munition, research into new technologies is probably a high risk business.

7. Waste Processing

Liquid effluent is processed by a company specialised in the destruction of industrial waste, which offers significant savings as it avoids investing in a dedicated unit. After precipitation, the arsenic salts are stored in old disused mines.

8. Conclusion

Destruction of old waste dating from the first faltering steps in chemical warfare and spread at random across forgotten battlefields is a long term, expensive task, requiring special equipment that is difficult to size and highly qualified staff. With prevailing environmental concerns, governments have a moral obligation to manage this activity correctly, respecting current regulations and ensuring protection for the population.

RECOVERED OLD ARSENICAL AND 'MUSTARD' MUNITIONS IN GERMANY

Technologies, Plans and Problems

HERMANN MARTENS
Wehrwissenschaftliches Institut far Schutztechnologien-ABC-Schutz
Postfach 1142
29633 Munster, Germany

1. Introduction

Among chemical warfare (CW) agents, the arsenicals do not exhibit high levels of acute toxicity. But because of the imperishability of arsenic as an elemental constituent of them, arsenicals jeopardize the environment in the long term. In comparison, 'mustard' agents may in principle be degraded to non-toxic products by the action of environmental conditions such as exposure to water, elevated temperatures and events that disperse it into fine droplets or particles.

The chemical munitions cleanup problem is however complicated by the fact that, during World War II, Germany manufactured special preparations of 'mustard' mixed with arsenic compounds of an exceedingly viscous and stable type, called "Zählost". Both constituents had insidious properties including long-term carcinogenic effects. Moreover, they pose problems as to their disposal whether by wet chemical (hydrolysis) or thermal processes.

The two classes of CW agents inherited from an unfortunate and belligerent past were unpleasant in their intended effects and in a different way are unpleasant when one undertakes destruction of them.

2. German World War I & II Arsenical and "Mustard Gas" Munitions

2.1 CHEMICAL STRUCTURE AND COMPOSITION

An imposing number of organoarsenic and 'mustard' compounds have been tried out for use as chemical agents. Many of them were rejected because they lacked toxicity or on account of tactical drawbacks. In the case of 'mustard'-type blister agents, the number was increased until 1945 by manifold tactical formulations. The arsenicals of German war production are listed in Table 1 [1] [2]. The basic "mustard gas" compounds may be found in Table 2 [1] [2] and tactical mixtures are summarized in Table 3 [2].

J.F. Bunnett and M. Mikołajczyk (eds.), Arsenic and Old Mustard: Chemical Problems in the Destruction of Old Arsenical and 'Mustard' Munitions, 33–78.

Actually some 'mustard' types contain considerable amounts of by-products arising from the production process. In particular, the "direct process" for synthesis of 'mustard' formed polysulfide by-products. If they were not separated in the course of fabrication, such polymer sediments in filled munitions become a problem during long-term storage. Also, problems may arise in the dismantling process owing to residues that obstinately stick to the walls of shells being emptied.

TABLE 1. Arsenicals of German war production

Chemical Designation	German Code	NATO/US Code
Methyldichloroarsine	Methyldick* Medikus	MD
Ethyldichloroarsine	Dick*	ED
Phenyldichlorarsine	Pfiffikus	PD
Diphenylchloroarsine Clark I	CI	DA
Diphenylcyanoarsine Clark II	CII Dora	DC
Diphenylaminochloroarsine (Adamsite)	Azin	DM
50%Phenyldichloroarsine 35%Diphenylchloroarsine 5%Triphenylarsine 5%Arsenic(III)-chloride	Arsinöl (A-öl)	—
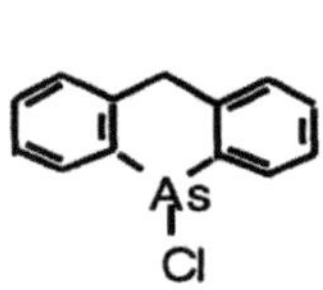	Excelsior	—

* World War 1 product.

A further complication is that substances were sometimes intentionally added to the original 'mustard' products in order to reduce the freezing point for winter use or to improve persistency after CW agent dispersal on the ground and on surfaces of military equipment.

TABLE 2. Mustard gas types of German war production

Designation/formula	**German name**	**German code**	**NATO/US code**
Sulfur mustard $S(CH_2CH_2Cl)_2$: CH_2CH_2Cl / S / CH_2CH_2Cl (made from thiodiethanol)	Oxolost S-Lost (summer mustard)	O, OL	HD
Sulfur mustard CH_2CH_2Cl / S / CH_2CH_2Cl (made from S_2Cl_2)	Direktlost S-Lost	D, DL	HD
Oxygen mustard $CH_2CH_2–S–CH_2CH_2Cl$ / O / $CH_2CH_2–S–CH_2CH_2Cl$	O-Lost	OB	T
Sesqui-mustard $S–CH_2CH_2Cl$ / CH_2 / CH_2 / $S–CH_2CH_2Cl$	Sesquilost	DO Doppel-O	Q
Propyl mustard $CH_2CH(Cl)CH_3$ / S / CH_2CH_2Cl	Propyllost		
Nitrogen mustard CH_2CH_2Cl / N–CH_2CH_2Cl / CH_2CH_2Cl	Stickstofflost N-Lost	T9 K	HN-3

Noteworthy in particular were German formulations of viscous 'mustard'[1] that were intended to confer long-term persistence on surfaces and to thwart efforts of decontamination. Like 'mustard' formulations for winter use, some of the Zählost types have arsenicals as major components. Thickening materials that were available up to 1945 (e. g. chlorinated rubber, waxes and polystyrene), added for powerful enhancement of viscosity, incidentally made the chemical agent formulations completely insoluble in the usual technically available solvents.

Thus, Zählost caused considerable technical disposal problems when, about 1975, it was necessary in Germany to find an adequate CW agent destruction process.

The numerous aformentioned mixtures of 'mustard' agents with organoarsenic compounds aggravated definitive identification by the use of qualitative and quantitative analysis.

'Mustard' agents are remarkably resistant to hydrolytic or other wet-chemical degradative influences when they are in bulk without intense contact with water phases. Some of the arsenical agents are solids (e.g. Adamsite, Clark I & II), poorly soluble in water, and consequently even more stable. Residues of Adamsite have been found on the ground surface of former munition filling stations or at loading sites, unaffected by degradation after more than five decades of exposure to wind and rain.

Most of the the arsenicals listed in Table 1 were produced and used on battlefields during World War I. Excelsior and arsine oil were products of World War II, during which the arsine oil mixture was utilized in large amounts to winter-proof the various types of "mustard gas".

Clark I & II and Adamsite are well known irritants that can be found in shells of both World Wars, Adamsite sometimes admixed with nitrocellulose flakes. Excelsior was considered to be superior to Adamsite as an irritant and reached the stage of pilot plant production. About 10 tons of Excelsior, and 50 tons of diphenylmethane-o-arsonic acid (Merodan), an intermediate in its manufacture, are still hidden deep underground in a former salt mine.

Owing to extended field testing in the Munster area during World War I and a violent explosion on October 24, 1919 that destroyed all the facilities for production and filling of chemical munitions as well as storage places for some hundred thousand chemical shells, this area received a burden of contamination by CW agents that remains substantial even unto this day. The situation was further aggravated by similar activities during World War II, following the inauguration of a new chemical warfare test center built in 1935-38. An infrastructure, including large, partially underground filling stations, finally was demolished in 1948, again with momentous dispersal of CW agents onto the terrain.

In view of the diversity of chemical fillings in an even larger diversity of munitions, it seems obvious that many analytical studies and much research on the fates of arsenicals under environmental conditions (soil and ground water) should be carried out. Needed also are similar investigations with respect to the different types of 'mustard'.

---

1 Editor's note: Zählost is pronounced "tsay–loast", approximately.

TABLE 3. German "mustard gas" formulations

Sulfur 'mustard' %	mixed with %	World war	German code
80	20 chlorobenzene	1	Lost
50	50 chlorobenzene	2	C
50	50 anthracene oil	2	OR (Winter 'mustard')
40-50	50 - 60 oxygen 'mustard'	2	OB
30	70 oxygen 'mustard'	2	OB
50	10 oxygen 'mustard' 40 propyl 'mustard'	2	OM,OKM (Winter 'mustard')
88	8 oxygen 'mustard' 4 sesqui 'mustard'	2	OL
54	25 phenydichlorarsine 21 diphenylchlorarsine (Arsinöl)	2	OA (Winter 'mustard')
	91 OA 6 Tornesit (chlorinated rubber) 3 S-wax	2	Z-OA 396 Zählost (viscous 'mustard')
	90 OM 7 polystyrene 3 S-wax	2	Z-OM 8

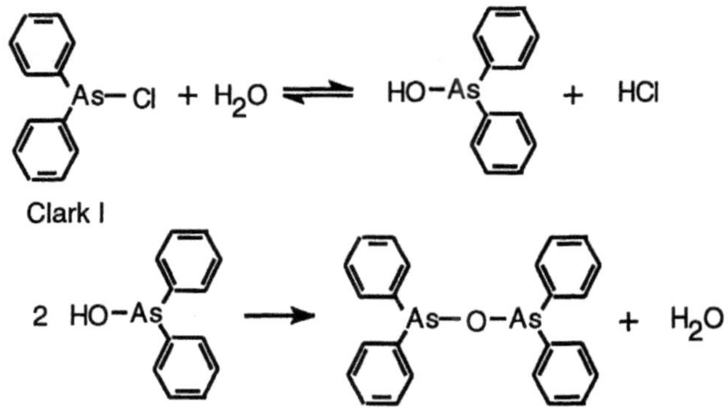

Figure 1. Hydrolytic degradation of Clark I.

Analytical work is much aggravated by the different states of preservation of the chemical agents inside and outside of the container walls and by the fact that the assortment includes chemical munitions captured from the Allies as well as German munitions. Nevertheless, because of other priorities, little basic research has been done. Analytical work has been mainly confined to determination of total arsenic, water soluble arsenic, and residual 'mustard' (i.e., that does not pour out).

However, Clark I that lay on the ground since the early 1940's, when field testing was conducted, was transformed into bis(diphenylarsine)oxide (Figure 1). Scientists in Munster succeeded in finding this compound as a well crystallized component of a soil sample. Needles were revealed by scanning electron microscopy.

Because of the lack of uniformity of the discovered and collected chemical munitions in Germany and particularly in Munster, an incineration process for destruction of their chemical fillings was chosen as the only reasonable yet economical means of disposal.

2.2 MUNITIONS WITH MIXED ARSENICAL AND "MUSTARD GAS" FILLINGS

When British occupation forces in 1945 took over the Munster area, they identified chemical munitions that the army and air force of the Third Reich kept in storage bunkers or in field testing facilities. The detailed descriptions of the measurements and markings of the chemical munitions by British experts [3] are still the basis for initial identification of items found unless the outer surface has been completely destroyed by rusting. They also made drawings of munition internal structures and identified CW and high explosive (HE) fillings as well as fuze types and other characteristics.

Available for issue to combat units during World War II were hand grenades, shells of calibre from 7.5 to 15 cm, rockets of calibre 15 cm, mortar bombs of calibre 10 cm, chemical mines with 10 l. filling of Zählost, and aircraft gas bombs of 250 kg and 50 kg weight filled with "mustard gas" or Adamsite. Adamsite/nitrocellulose and Clark II/nitrocellulose mixtures also were available in thermal generators (smoulder pots) with about 25 kg fillings. In Figures 2-10, drawings of the most common chemical shells that had "mustard gas" and arsenical fillings, taken from the British report of 1945 [3] on the Munsterlager installations, are shown.

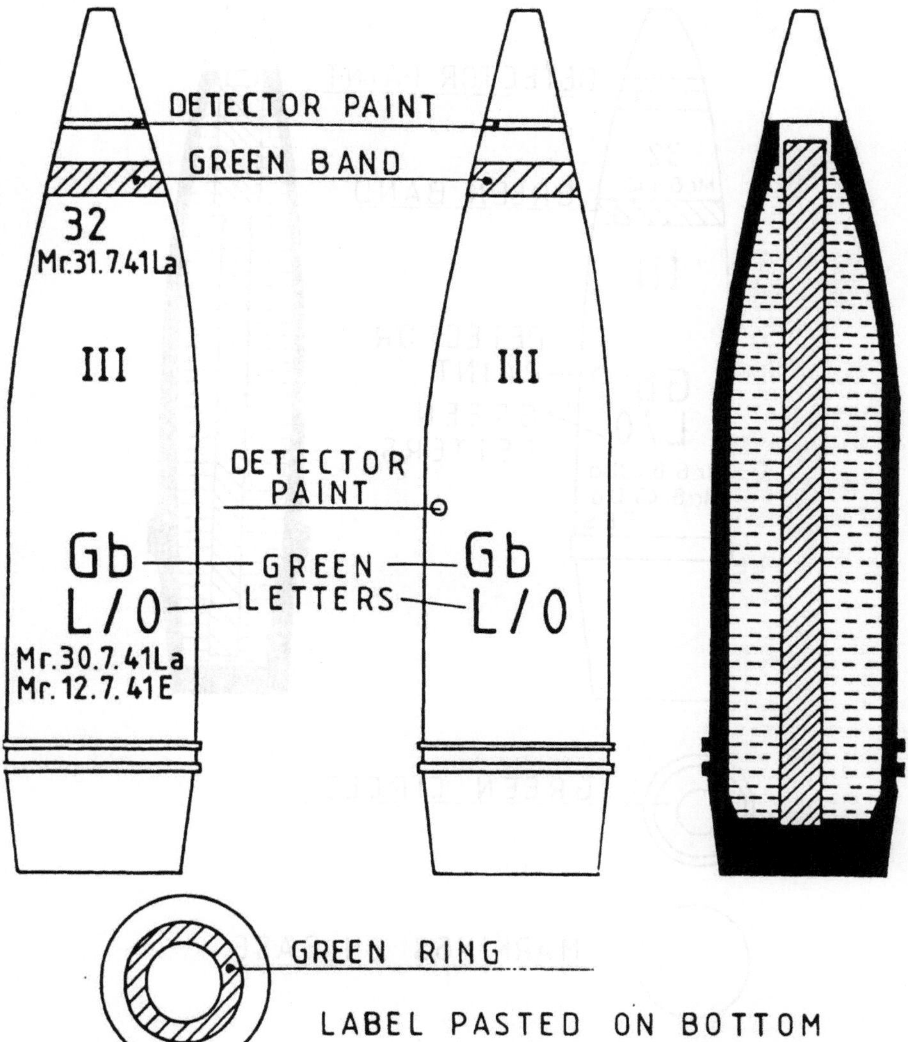

WEIGHT	36.8 kg
FUZE	AZ 23 Nb
CW FILLING	3500 ml O (HD)
HE FILLING	580g PETN/WAX 70/30

Figure 2. 15 cm Shell Grünring

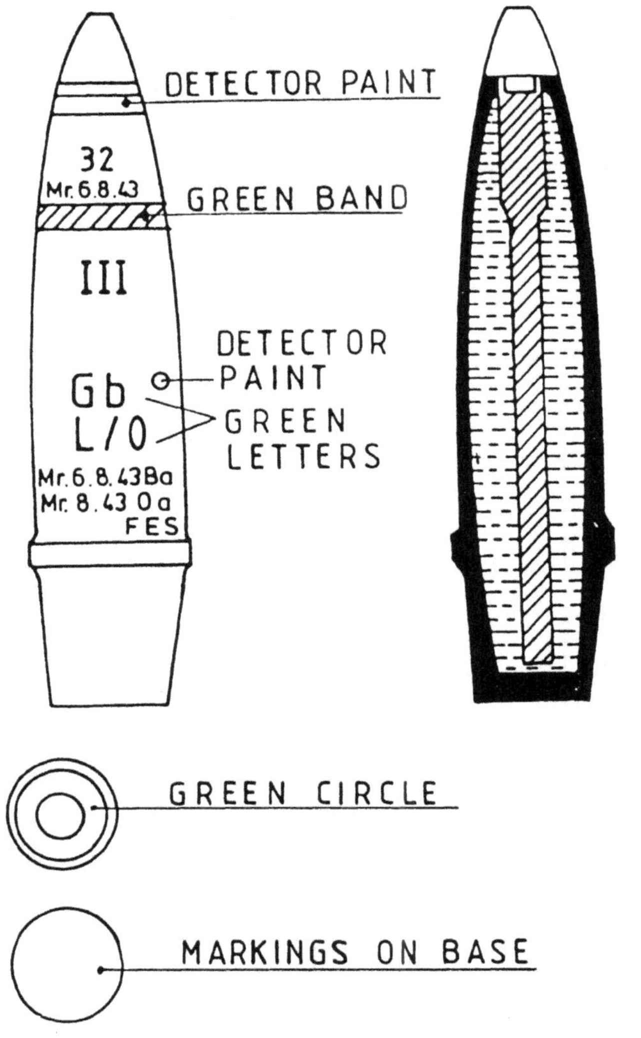

WEIGHT	13.7 kg
FUZE	Kl AZ 23 Nb
CW FILLING	1250 ml WINTERLOST
HE FILLING	125 g PETN/WAX 95/5

Figure 3. 10.5 cm Shell Grünring (field howitzer)

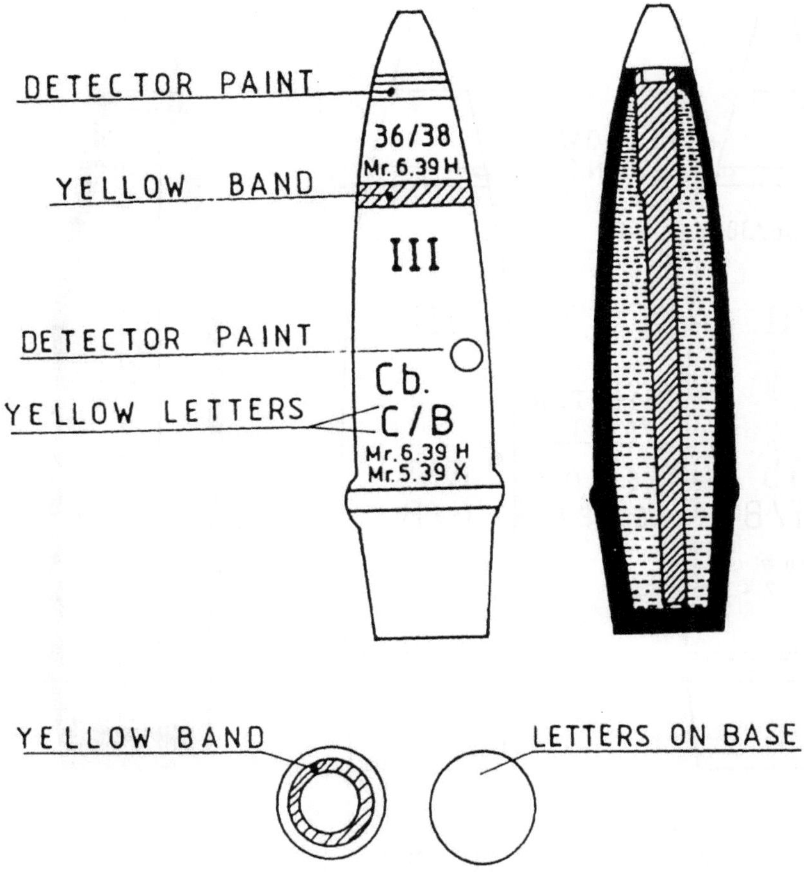

WEIGHT	14.0 kg
FUZE	Kl AZ 23 Nb
CW FILLING	1250 ml WINTERLOST
HE FILLING	94 g PETN/WAX 60/40

Figure 4. 10.5 cm Shell Gelbring (field howitzer)

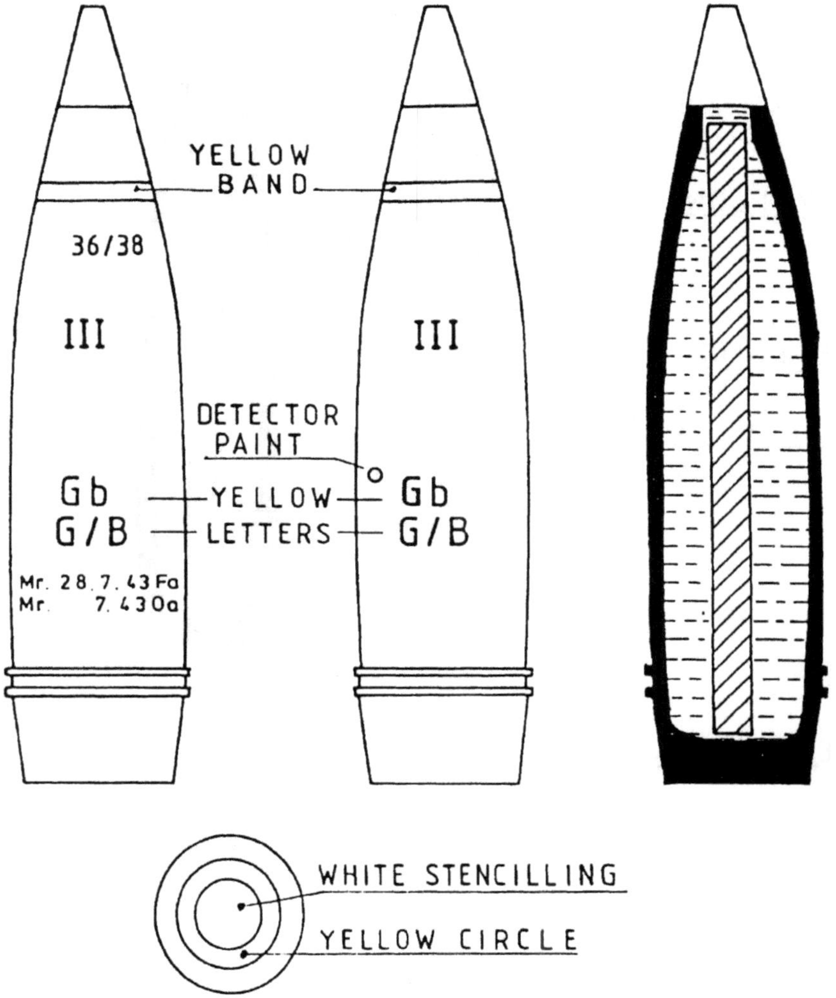

WEIGHT	37.4 kg
FUZE	AZ 23 Nb
CW FILLING	3500 ml (4.3 kg) WINTERLOST
BURSTER	580 g PETN/WAX 60/70

Figure 5. 15 cm Shell 19 Gelbring

There were four major designs of artillery muntions:

(a) A long shell with an inner burster to give initial cloud effects (Figure 2)

(b) A shell with a large head burster to achieve ground contamination (Figure 6)

(c) A shell charged with solid CW agent, e.g. Adamsite, embedded in a layer of high explosive (Figure 9)

(d) A shell having a short burster container to give extremely persistent ground contamination (Figure 12)

At first picric acid was used as burster filling for ground contamination and PETN for cloud effects. Later the picric acid burster was replaced by PETN-Wax 60/40 and still later the bursters were filled in three increments, the first and third being PETN-Wax 60/40 and the second PETN-Wax 35/65. A problem with the head burster version was loss by crater formation. An "improved" ground contamination effect was obtained by charging thickened 'mustard' into the shell. "Nitrogen mustard" was also tried for ground contamination but was found to be less effective than "sulphur mustard".

Trials in 1914-18 of a shell design (Figure 11) having Clark II in glass bottles embedded in high explosive showed poor performance because most of the CW agent was dispersed in relatively large particles; the generation of aerosols was desired. For this reason, use was made of a new design in which the solid material (Adamsite or Clark I) was arranged as a central cylinder or cone in paper carton, this being surrounded by high explosive (Figure 9).

The next stage in the development of arsenical shells was investigation of intimate mixtures of Adamsite with energetic materials, to function by thermal generation. After various trials with KNO_3, NH_4NO_3 and $KC1O_4$, it was concluded that a 50/50 mixture of Adamsite with nitrocellulose, with 1-2 % of urea as stabilizer, suffered the lowest losses by decomposition. Figure 10 shows the inner structure of a base ejection smoke shell.

Today the mixtures of Adamsite and nitrocellulose, and of Clark II with this type of gunpowder, present a severe problem to German chemical agent disposal experts. There is apprehension that the long-term storage of such mixtures has made them unstable, prone to explode, owing to nitro groups of the material. Mixtures of similar character were prepared for the generation of irritant smoke (aerosols) in smoulder pots or even larger thermal smoke generators. In these cases the nitrocellulose was plasticized with added alcohol and acetone and, after mixing with an equal weight of Adamsite, pressed together to form smoulder bodies. A specific microstructure of the mixed components was needed to guarantee its ignition as well as sufficient stability at high temperatures and a uniform velocity of burning with efficient smoke generation.

Unfortunately the aggregates changed in the course of years to such a hardness that it would be dangerous to crush them to a small particle size, as desired for processing in the course of CW destruction.

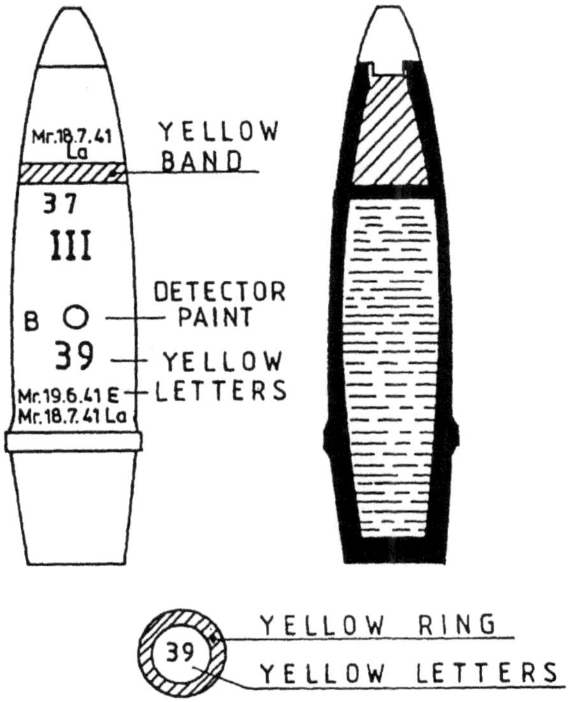

WEIGHT	13.3 kg
FUZE	Kl AZ 23 Nb
CW FILLING	1160 ml MUSTARD-ARSINÖL (WINTERLOST)
HE FILLING	208 g PETN/WAX 60/40

Figure 6. 10.5 cm Shell Gelbring (field howitzer)

A particular type of chemical munition is represented by the so-called "Sprühbüchse 37" (10 l. SpBü). This spray canister was intended to be used as a land-mine for ground contamination. A longitudinal cut-through is shown as Figure 12.

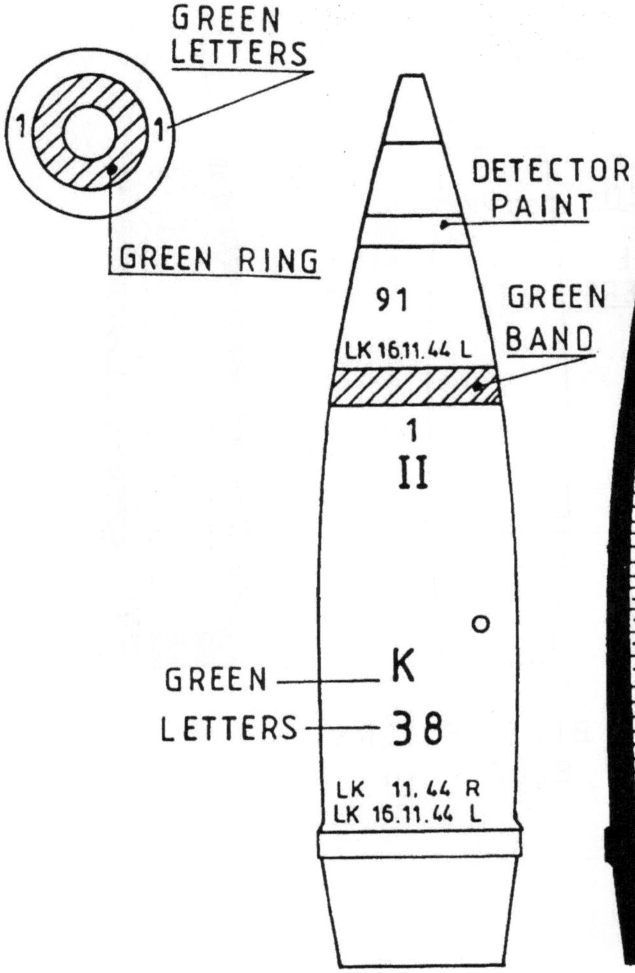

WEIGHT	38.1 kg
FUZE	Kl AZ 40 Nb
CW FILLING	2.9 kg HN-3 (98% PURE)
HE FILLING	2.1 kg RDX/WAX 95/5

Figure 7. 15 cm Shell 38 Grünring 1

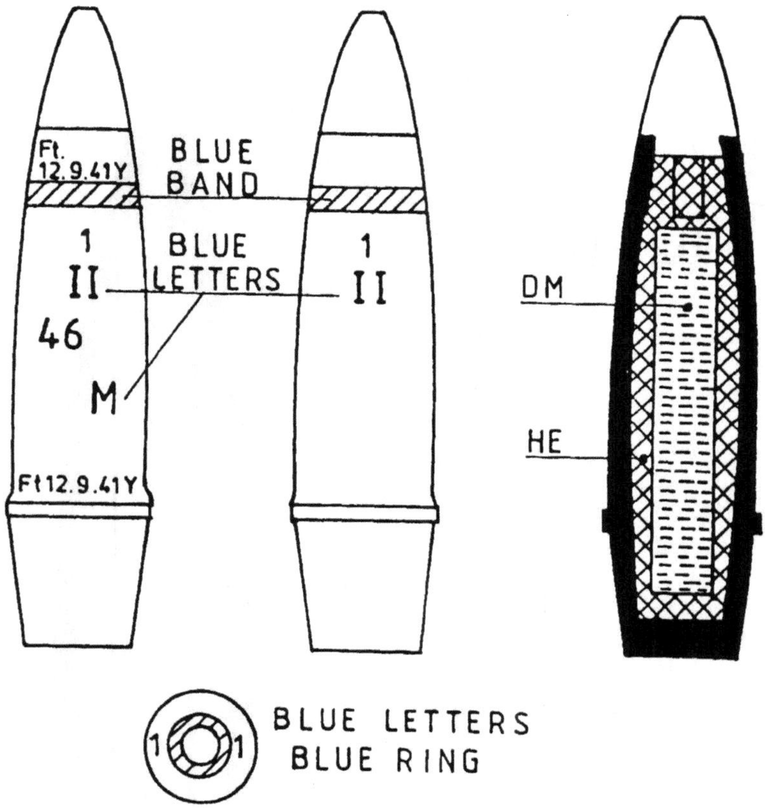

WEIGHT	14.8 kg
FUZE	AZ 23 umg 0.15
CW FILLING	550 g DM (90% PURE)
HE FILLING	900 g TNT

Figure 8. 10.5 cm Shell Blauring 1 (field howitzer)

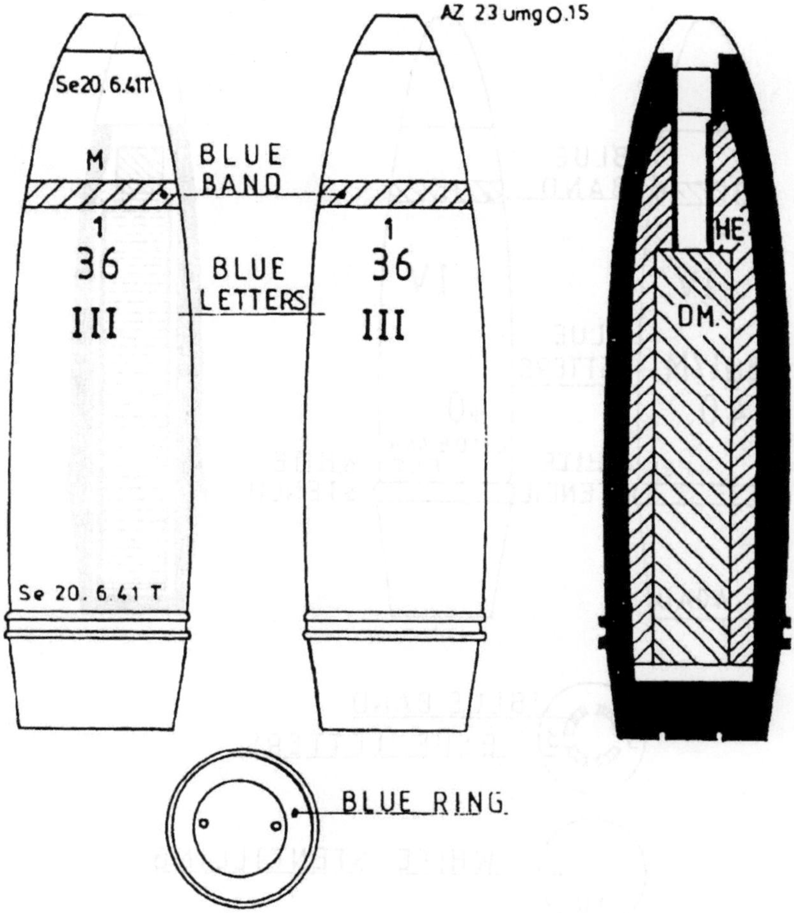

WEIGHT	42.3 kg
FUZE	AZ 23 umg 0.15
CW FILLING	1485 g DM (90% PURE)
HE FILLING	2338 g PETN/WAX 75/25

Figure 9. 15 cm Shell 19 Blauring 1

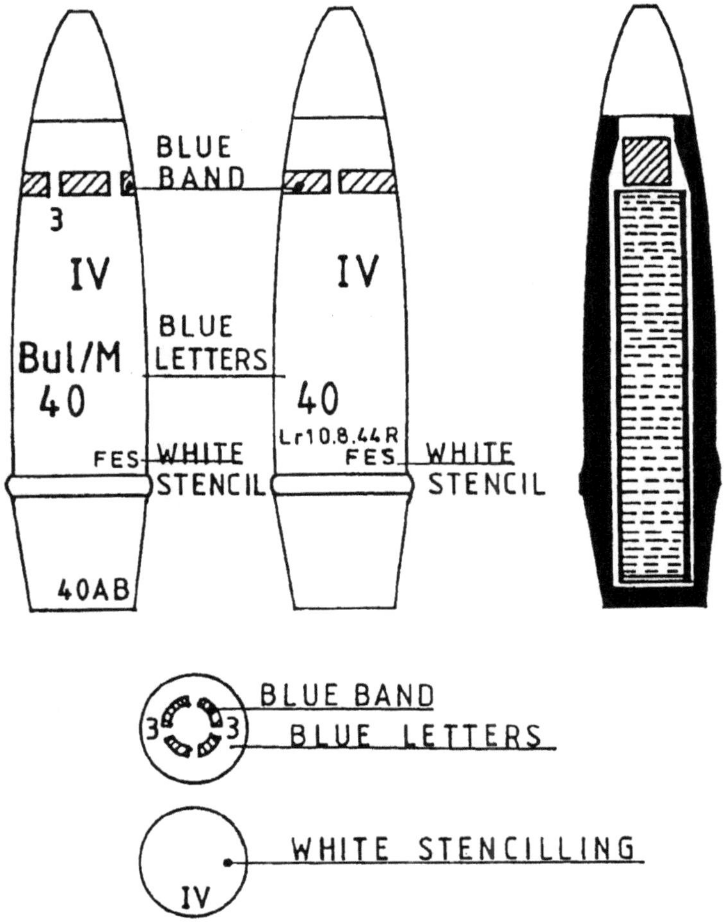

WEIGHT	14.1 kg
FUZE	DOPP 25/60 Flor 25/60V
CW FILLING	837 g DM/NITROCELLULOSE
EJECTION CHARGE	84g COMPRESSED GUNPOWDER
TIME OF EMISSION	1-2 min

Figure 10. 10.5 cm Shell 40 Blauring 3 (field howitzer)

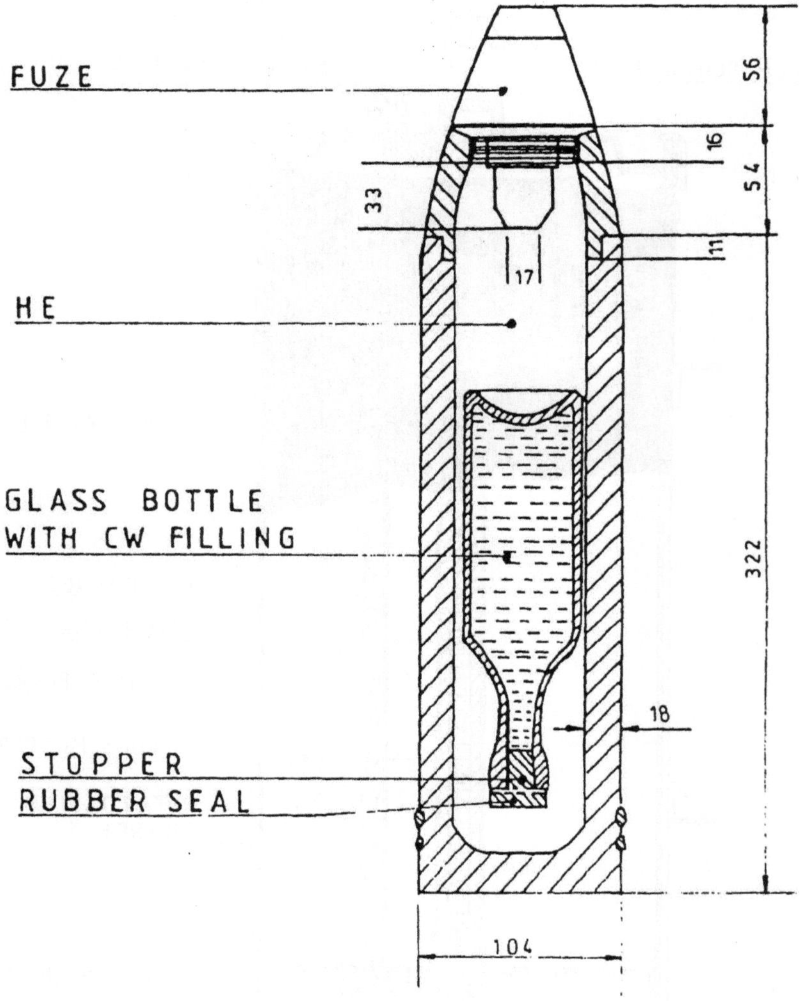

FUZE	PICRIC ACID
CW FILLING	CLARK II

Figure 11. 10.5 cm Shell Blaukreuz 2 (World War I)

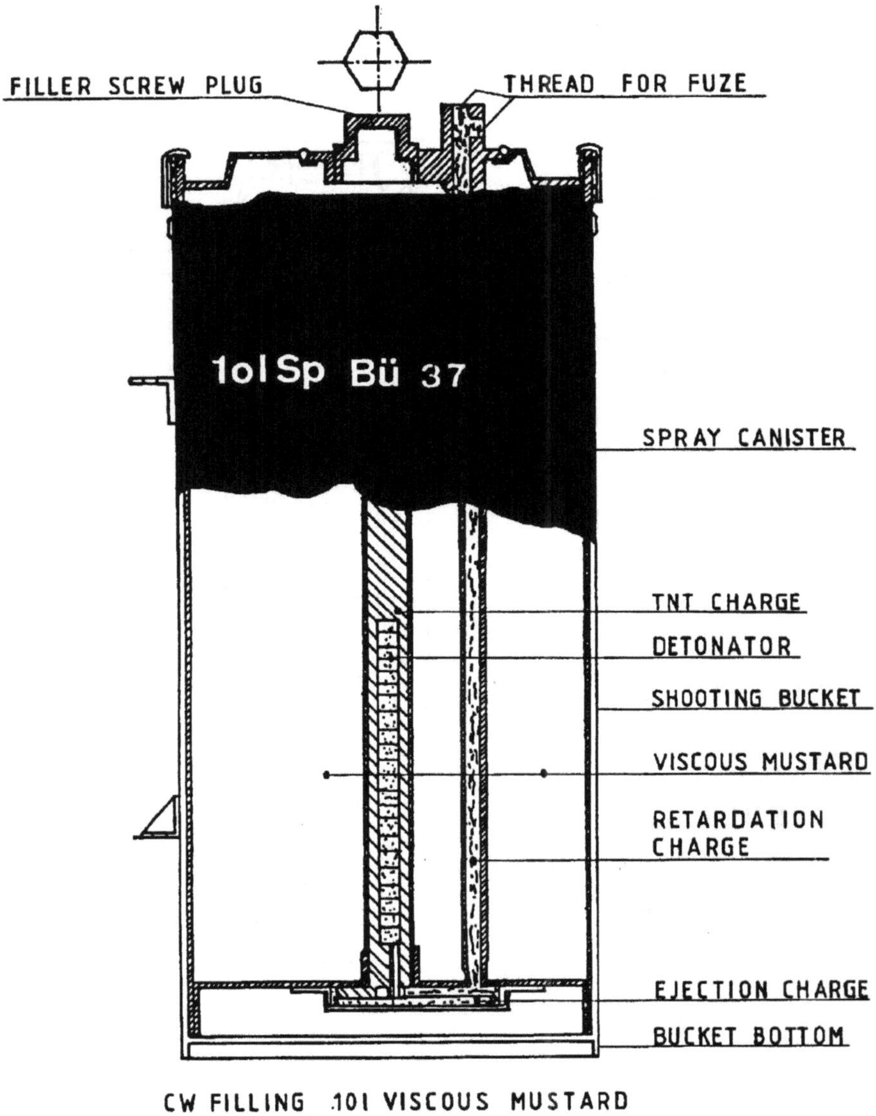

Figure 12. German Ten Liter Spray Canister

The 10 l. SpBü contained 10 liters of highly viscous (some hundred poise) Zählost, with a rather broad variation in sort and percentage of constituents. Most of the formulations were combinations of 'mustard' with arsine oil (called "winter mustard") with the addition of thickeners. The principal thickeners were chlorinated rubber (Tornesit), wax (S-Wachs) and polystryrene. The ejection charge blew the container with the Zählost to an average height of 20-30 m, where it split apart. The resulting agent dissemination contaminated a ground area approximately 50 m in diameter. The purpose of this extremely sticky CW agent was to blockade terrain against the passage of enemy troops. Its main feature was that no decontamination activity except burning it would have been able to destroy the Zählost.

Field testing of the application of Zählost during World War II heavily contaminated some sites. At that time there was no way of cleanup other than deep ploughing. Small aggregates of Zählost, globules encapsulated by the polymer thickeners, have endured over 50 years. In some soil samples taken from the ground of a former testing range, “mustard gas” could be determined by gas chromatography. It is known that on the bottom of the Baltic Sea much larger aggregates of Zählost, mostly in lumps, are still present. The encapsulated Zählost is still an effective blister agent when it contacts the human skin. Burns on the hands of Danish fishermen were reported only a few years ago. After almost total corrosion of the metal parts of the spray canisters in sea water, the highly viscous CW agent fillings nevertheless endured owing to the formation of external protection layers of resinous thickener material.

Arsenicals and various types of 'mustard' agents were also filled into aircraft bombs, mortar bombs (Figure 13) and hand grenades. The KC 250 bomb type (Figure 14) was charged with 100 kg of ("sulphur mustard") and had a burster charge of 3.2 kg of TNT. It was said effectively to contaminate an elliptical ground area of 600 m^2. A 50 kg bomb holding 13 kg of 50/50 Adamsite/nitrocellulose (Figure 15) acted, after hitting earth, as a static generator for irritant smoke. In order to impede ground penetration by the munition, the terminal velocity was reduced by a drag plate secured across the tail fins. The emission period was about 4 minutes, during which it generated toxic concentraions of 500-1000 mg/m^3 at a distance of 25 m. Even 10 km away, CW agent at a level of minimum effects (about 0.1 mg/m^3) was detectable.

Markings on the outside of the munitions were applied for easier recognition. The most informative markings of the second World War were coloured rings (Table 4). They indicated the physiological effect of the charge as dependent on the form (agent cloud or liquid contamination) in which it was dispersed:

Yellow	Vesicant (blister effect)
Green	Choking or systemic effect
Blue	Sternutator (sneeze agent)
White	Lachrymator (“tear gas”)

Numbers associated with the rings gave information of different chargings.

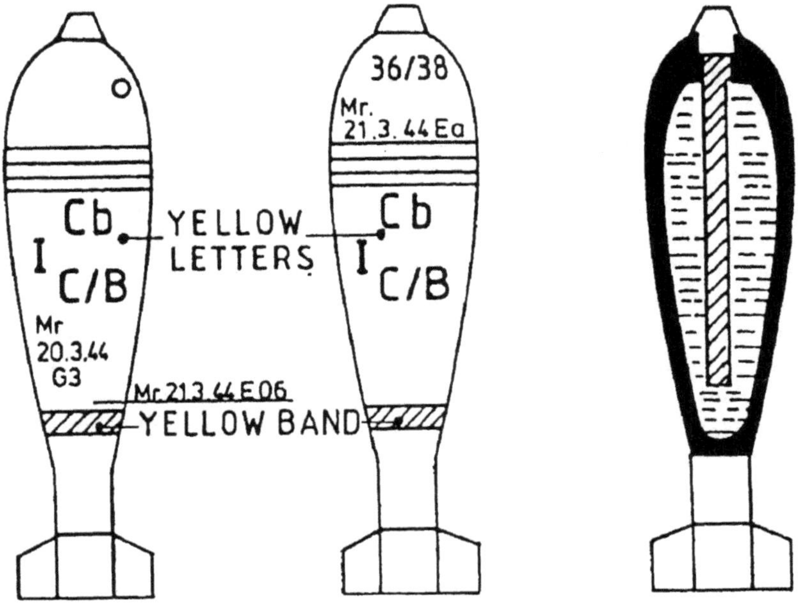

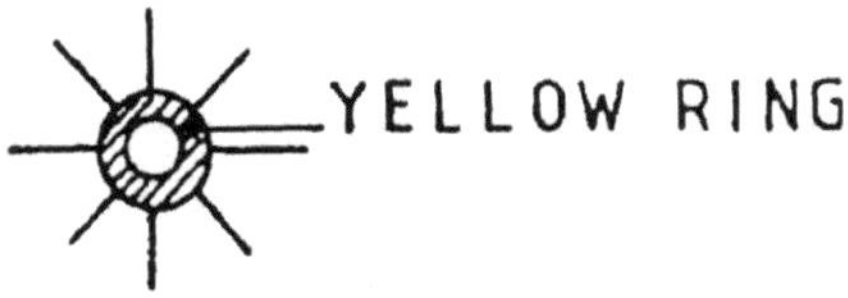

WEIGHT	7.0 kg
FUZE	Wgr Z 38
CW FILLING	1080 ml WINTER LOST
HE FILLING	50 g PETN/WAX 60/40

Figure 13. 10 cm Mortar Bomb 35St Gelbring

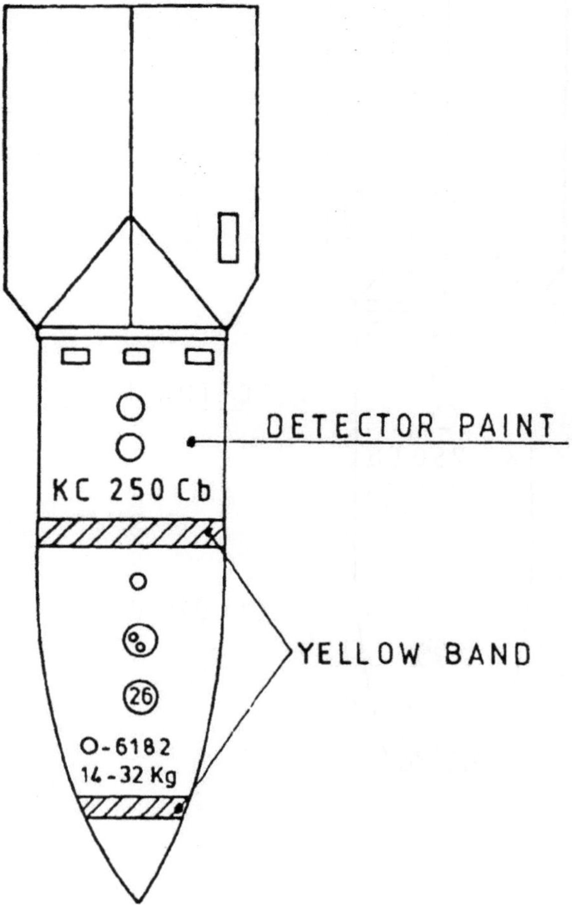

WEIGHT	160 kg
FUZE	El AZ (26)
CW FILLING	100 kg 0 (HD)
HE FILLING	3.2 kg TNT

Figure 14. Aircraft Bomb KC 250 Gb

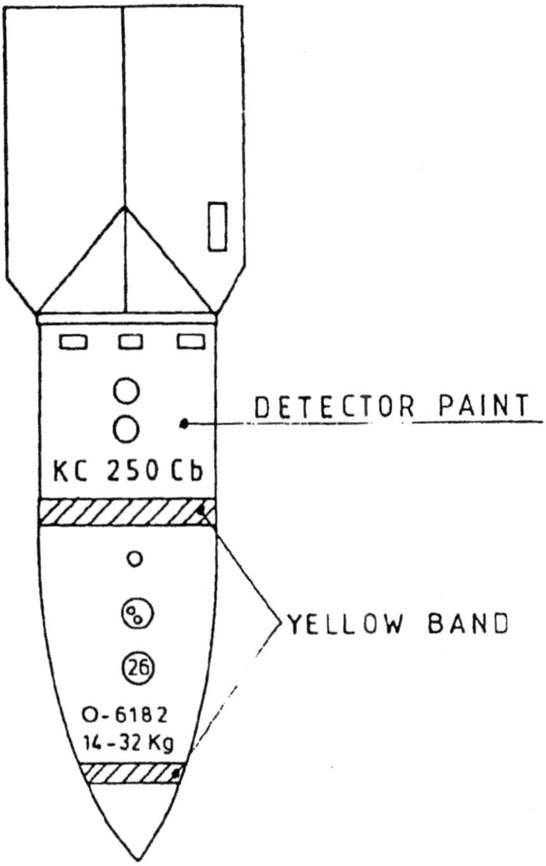

WEIGHT	43 kg
FUZE	EL AZ (46)
	14 kg DM/NITROCELLULOSE 50/50
TIME OF EMISSION	4-6 min

Figure 15. Aircraft Bomb KC 50 II Bu

Letters, numbers and coloured rings were painted, stenciled or stamped on the field-grey surface paint. Because of corrosion, such markings have disappeared in the course of elapsed decades. For this reason non-destructive techniques like X-ray photography and neutron activation analysis are used when possible for munition identification.

TABLE 4. Markings on German World War 2 CW munitions

Yellow ring	"mustard gas", small burster
Double yellow ring	Thickened mustard gas, small burster
Green ring	"mustard gas", medium burster
Green ring 1	'Nitrogen mustard', large burster
Green ring 3	Tabun, head burster
Green ring yellow	"mustard gas", large burster
Blue ring 1	Adamsite with exterior burster
Blue ring 2	Clark in arsine oil, central burster
Blue ring 3	Adamsite, base ejection generator
White ring	Chloroacetophenone/high explosive intimate mixture

3. Manipulation of CW Munitions, in Particular "Mustard Gas" and Arsenical Munitions

3.1 RECOVERY AND DISMANTLING

German old chemical munitions still lie hidden in the ground around Munster, Lower Saxony. Other sites of similar character on German territory are known, such as Hallschlag in Rhineland-Palatinate and Löcknitz in Mecklenburg-Vorpommern (not far from Szczecin, Poland), but none is as densely contaminated as the area near Munster.

Likewise some smaller storage facilities exist in the federal states, but the storage bunkers (Stradley type) in Munster are much larger and are built according to the latest technical safety standards.

Before dismantling, which means mechanical opening of the munitions followed by emptying from them the CW agent fillings with separation of high explosives and fuzes, the following steps normally have to be executed:

Reconnaissance and location in the terrain
Unearthing and type identification
Removal and transportation
Intermediate storage
Preparation for dismantling:
- X-raying
- Non-destructive methods for identification of CW filling

The next step, dismantling, governs the speed of the whole disposal process. Dismantling is carried out in Munster in a specially hardened building with separate chambers for remotely controlled mechanical operations like drilling, sawing and planing of munition items.

For effective disposal of Zählost, the German CW agent incineration plant at Munster was designed as a batch-type double chamber incinerator.

Zählost can be removed from the casings (spray canisters) after the lids have been cut off only by manual work, an extremely hazardous activity. Decontamination of the disassembled parts and of the tools used, like spoons and spatulas, is possible only by using the very effective German emulsion, which contains tetrachloroethylene as solvent and calcium hypochlorite as oxidant. The used decontaminant solutions constitute a further disposal problem, because they cannot be released into the waste water channel but must be burned, like liquid CW agents, in the incineration plant in Munster.

Solid material, e.g. Clark or Adamsite, can be removed from the cylindrical part of a shell by using a push-out piston after the conical top and bottom of the munition have been cut off. Self-evidently, it often happens that arsenicals become mixed with high explosive of the demolition charge. Such occurrences necessitated planning and building a second incineration plant in Munster because solid agents cannot be burned economically in the first incinerator. Moreover, mixtures with high explosives are dangerous to burn unless first submitted to a crush-down and attenuation treatment. After separation in the dismantling steps, explosive charges and fuzes are still blown up at an open-air demolition site. Development work is under way at some German companies to construct furnaces for controlled high explosive incineration.

In the context of this paper, only methods for CW agent identification in munitions containing arsenicals and Zählost are described in detail.

3.2 CHEMICAL AGENT IDENTIFICATION

After study of the outside characteristics of a found munition, the first step for identification of the filling is X-raying its inner structure. This can be executed at the site of discovery by means of mobile X-raying devices. On site X-ray examination is essential if the item is not transportable because of explosion or CW agent spill hazard. In most cases munitions are removed and transported to an intermediate storage place until being sorted out for dismantling.

Very few chemical munitions can be identified by their outside appearance and visual markings. Often samples must be taken and examined by analytical chemistry or physical methods. There are no great problems in carrying out GC/MS, atomic absorption or IR-spectrometric analyses in case a well equipped laboratory is nearby. Chemical identification however requires hazardous mechanical intervention, such as drilling a hole into the munition or cutting through the wall to take samples of the filling. For this reason each item is initially X-rayed to classify it according to its inner structure and find out how it can be safely opened and dismantled.

Although chemical laboratory techniques cannot be renounced for final confirmation of the chemical nature of the CW filling, non-destructive methods enable easier and faster recognition of the CW agent.

3.3 NON-DESTRUCTIVE EVALUATION (NDE) OF CHEMICAL MUNITIONS

Besides the light elements hydrogen (H), carbon (C) and sometimes oxygen (O), all chemical warfare agents contain one or more of the following characteristic elements: Fluorine (F); Chlorine (Cl); Bromine (Br); Sulphur (S); Phosphorus (P); Arsenic (As).

To a considerable extent, determination of which of these elements is/are present serves to characterize the agent. In the course of preparations to implement the Chemical Weapons Convention, physical methods to detect these characteristic elements have been developed for identification of the chemical fillings of CW munitions without opening and taking samples. In the German Armed Forces Scientific Institute for Protection Technologies-NBC-Protection in Munster, five methods are being used or are in development:

(1) Neutron activation analysis (NAA)
(2) Neutron induced prompt photon spectroscopy (NIPPS)
(3) Hydrogen concentration measurement (HCM)
(4) Acoustic resonance spectroscopy (ARS) using ZONIC's world class analyser (WCA)
(5) Ultrasonic pulse echo (UPE).

The function and capabilities of these methods are now briefly described.

3.3.1 Neutron Activation Analysis

Neutron activation analysis (NAA) has been used for many years with stationary equipment for identification and determination of chemical elements. Of present interest is its use to determine the three characteristic elements chlorine, arsenic and bromine, which appear in arsenicals and some blister agents. Efforts are under way now to develop mobile equipment.

As can be seen from Figure 16, NAA is a two step process. In the first step the sample, e.g. a shell with a chemical filling, is irradiated with neutrons (thermal or fast neutrons from a Cf-252 neutron source). This neutron irradiation generates radioactive isotopes in the sample. Most of the isotopes emit γ-rays with half-lives characteristic for the respective isotope. In the second step these γ-rays are measured and analysed with a high resolution γ-ray spectrometer with hyperpure germanium (HPGe)-detector. The nuclear reactions involved and characteristics of their γ-ray emission are shown in Table 5.

TABLE 5. Selected thermal neutron capture reactions and γ-ray energies of the isotopes produced

Element	Nuclear reaction	Half life	γ-Energies [keV]	Relative intensity
Arsenic	$^{75}As\ (n,\gamma)\ ^{76}As$	26.4 h	559	44.6
			657	6.4
Chlorine	$^{37}Cl\ (n,\gamma)\ ^{38}Cl$	37.2 min	1642	32.8
			2167	44.0
Bromine	$^{79}Br(n,\gamma)\ ^{80}Br$	17.6 min	617	7.2
			666	1.1

Results from this non-destructive method may be expected after a measuring time of 15-30 minutes.

3.3.2 Neutron Induced Prompt Photon Spectroscopy (NIPPS)

Neutron induced prompt photon spectroscopy (NIPPS), in principle, permits determination of all elements that appear in chemical warfare agents, high explosives, smokes and incendiaries. The elements emit photons of different characteristic energies after inelastic encounter with a neutron, or neutron capture. These photons can be detected with an HPGe detector (see Figure 17), with high resolution. The necessary measuring times are between 1 and 30 minutes, depending on the amount of agent inside the shell and on the steel wall thickness. In this case, on-line element identification by a computer connected to a multi-channel analyzer detection system is possible; the algorithm makes use of a suggestion as to which CW agent is probably present.

The relevant nuclear reactions for NIPPS are the inelastic scattering of fast neutrons, e.g. ^{19}F (n, n', γ) ^{19}F, or the capture of thermal neutrons, e.g. ^{35}Cl (n, γ) ^{36}Cl. Hydrogen, chlorine and arsenic can be detected with high sensitivity and without any problems.

3.3.3 Hydrogen Concentration Measurement

Hydrogen concentration measurement (HCM) is a convenient method to carry out screening missions in depots with a variety of stored munitions, particularly to differentiate between conventional and chemical weapons. CW agents and explosives are differentiated according to their contents of hydrogen (H-atoms per cm^3), which actually vary considerably between the molecules of chemical agents (classical and modern) and high explosives like TNT, PETN and RDX.

In this method, fast neutrons emitted from an isotopic neutron source are used to irradiate the chemical filling of a munition. The hydrogen atoms of the chemical agent or explosive act as moderators and slow down the neutrons. The neutrons collide elastically with protons, which form the nuclei of the hydrogen atoms, with resulting loss of kinetic energy by the neutron. After several such collisions, a neutron becomes "thermal". Such neutrons are called thermal because they are in thermal equilibrium with their vicinity. On the average, thermal neutrons move as slowly as the protons of the surrounding molecules. They can penetrate the walls of a munition and be determined by a special Li-detection system.

The principle of HCM is illustrated on Figure 18. This figure also shows the schematic structure of a portable HCM unit intended to be used for on-site inspections in munition depots.

3.3.4 Acoustic Resonance Spectroscopy

In acoustic resonance spectroscopy, a munition is exposed toa beating with a hammer-type excitation tool. The transmitted energy causes the munition to oscillate with ground and overtone vibrations that are characteristic for the respective munition geometry. The intensities of the vibrations are measured with a piezoelectric vibration receiver attached to the examined object. From the ratio between the measured vibration spectrum and the excitation energy spectrum an acoustic resonance spectrum (ARS) can be derived. Munitions of the same type (and filling) have the same ARS.

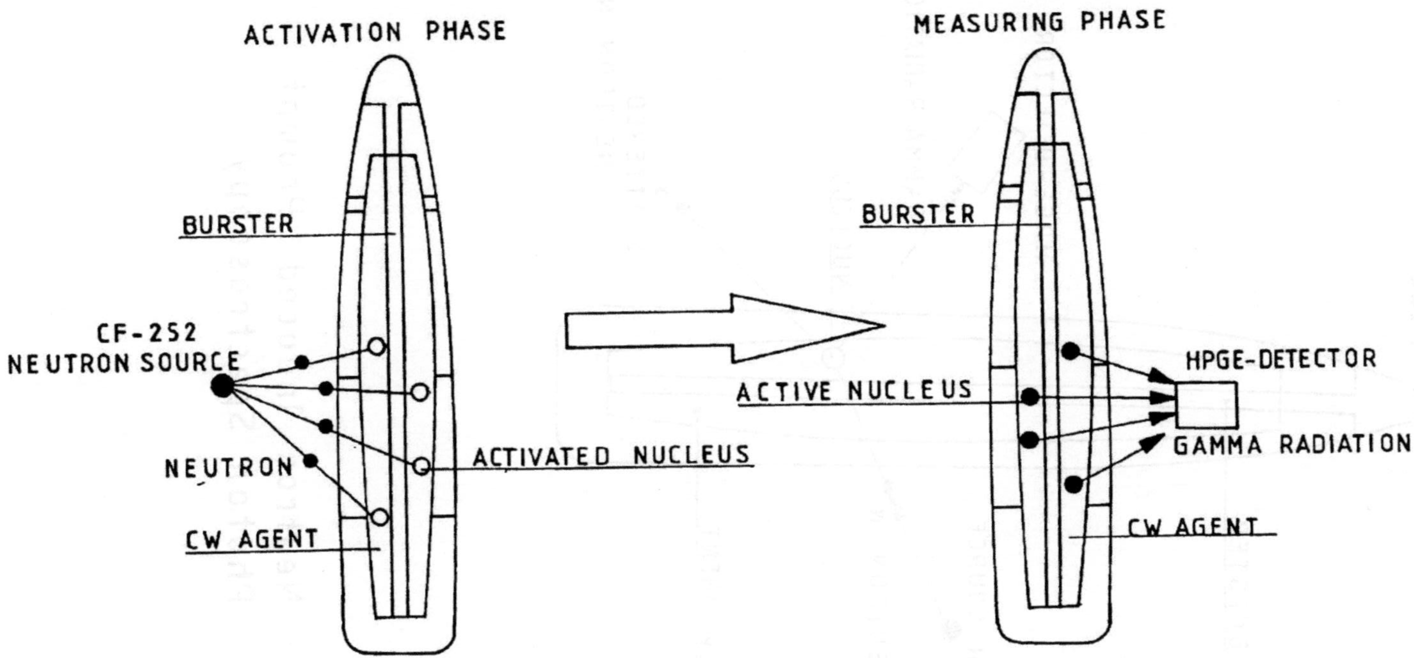

Figure 16 : Neutron Activation Analysis

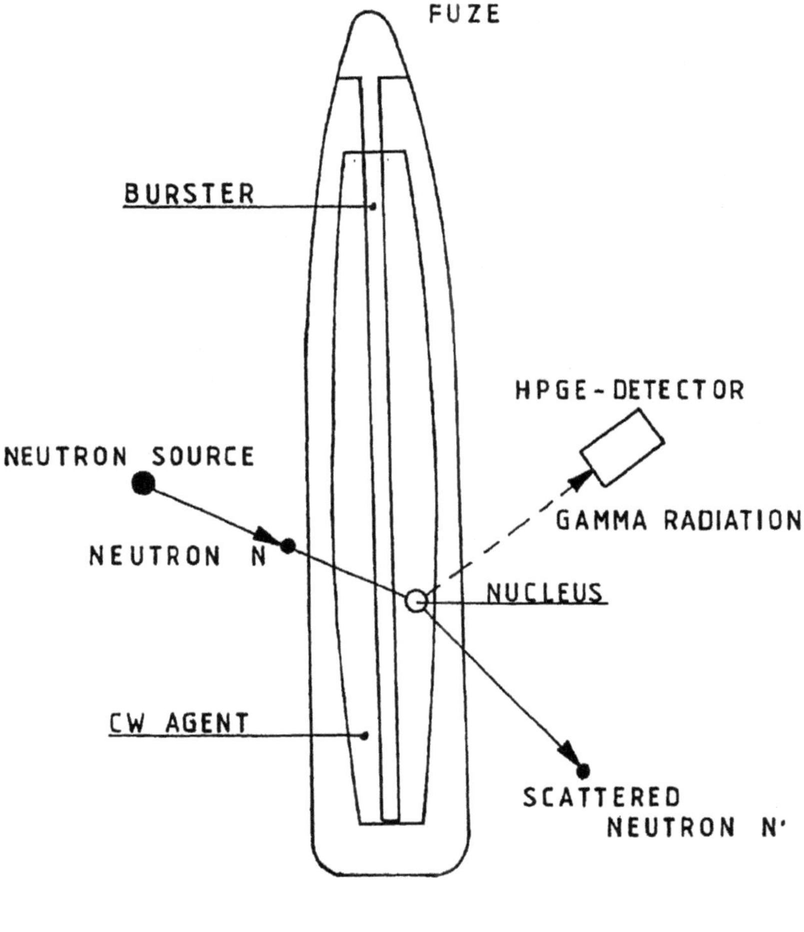

Figure 17: Neutron Induced Prompt Photon Spectroscopy

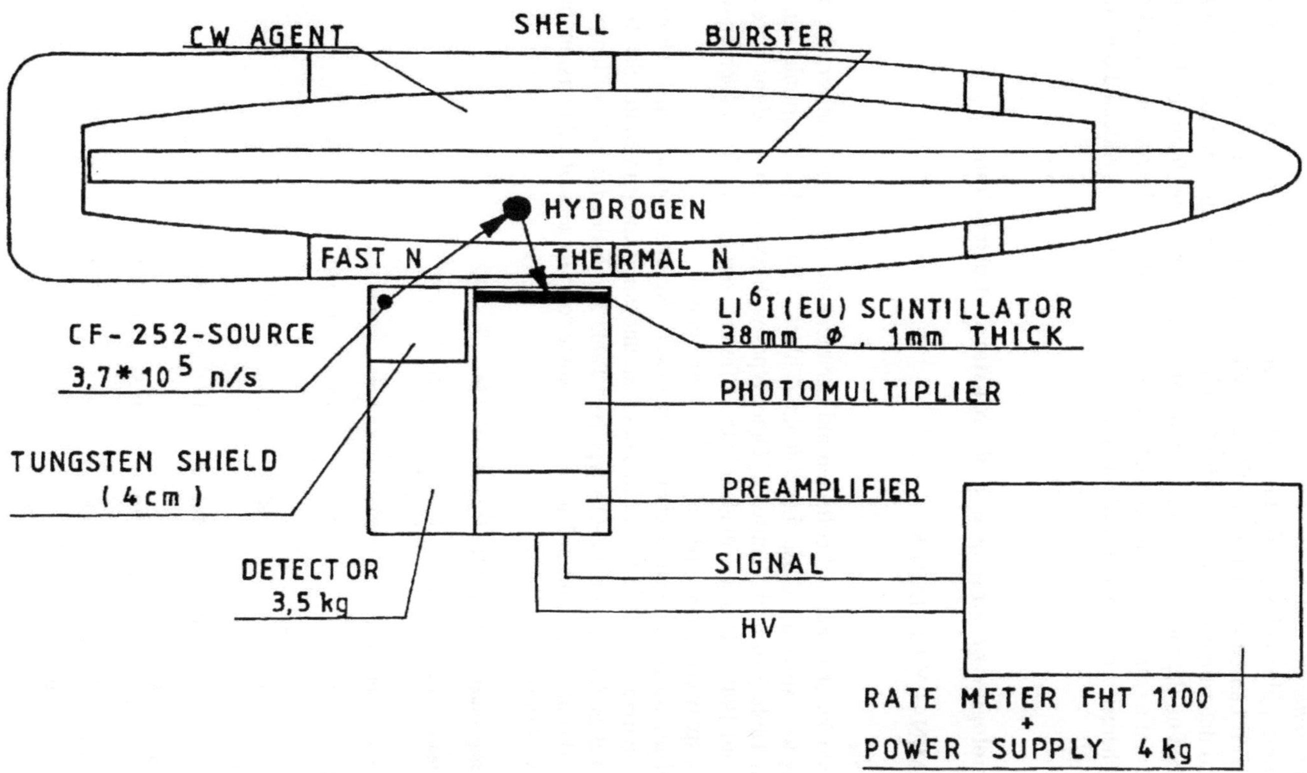

Figure 18: Hydrogen Concentration Measurement

3.3.5 Ultrasonic Pulse Echo

The ultrasonic pulse echo (UPE) method is used to measure the wall thickness of vessels. UPE can be adapted to distinguish chemical munitions from conventional types and to determine the filling level of shells. Differentiation between munition types is based on the fact that, besides the normal reflexion signal from the front wall, the detection of a rear-wall or burster-tube echo clearly indicates the presence of a liquid as filling, whereas with conventional munitions the original pulse is reflected only from the front wall.

4. Technologies of Arsenical and 'Mustard' Destruction

4.1 PRELIMINARY REMARKS

4.1.1 Zählost

Of all the stocks of found and collected old chemical munitions in Munster, the most challenging to destroy are those that contain Zählost. Two Zählost formulations are recorded in Table 3 (last two entries). Other types with other compositions were also prepared, until 1945. This was evident in the different colourings of the materials when a number of spray containers (10 l. SpBü 37) were opened.

Differences are also shown by chemical analyses. Results from a Danish laboratory [4] are summarized in Table 6. In this case, an undesired catch in Baltic Sea fishing was identified as Zählost of the Z-OA type (see Table 3). Inside the Zählost lumps was more than 40% of active 'mustard' agent, after some four decades of submersion, but in the "skin" was only about 0.1%.

TABLE 6. Composition of Zählost Z-OA from the Baltic Sea

Chemical formula	Name	Molecular weight	Mass %	k mol
$C_6H_5AsCl_2$	Pfiffikus	223	36.4	0.1632
$[C_6H_5]_2AsCl$	Clark I	265	4.5	0.0170
$[C_6H_5]_3As$	Triphenyl-arsine	306	2.7	0.0088
$AsCl_3$	Arsenic trichloride	181	1.8	0.0099
$[ClCH_2CH_2]_2S$	Mustard OL	159	45.5	0.2862
$[CH_2]_n$	S-wax	14n	9	0.6429
$C_6H_3N_3O_7$	Picric acid	229	0.1	0.0004

Besides the main constituents of Z-OA, arsenic chloride and triphenylarsine, both constituents of arsine oil, were identified. Probably dithiane was also present; it causes the characteristic smell of 'mustard' products.

The analytical results of the Danes demonstrate very clearly that any contacts of the skin with such catches from the Baltic Sea are hazardous to fishermen. Additionally, evaporation of 'mustard' and of arsenicals during warm or hot summer days creates a vapour hazard, dangerous to the respiratory tract.

One of the most annoying features of Zählost formulations was their almost total insolubility in solvents, owing to the thickening materials, in particular chlorinated rubber. This insolubility prevented direct dosage of the materials in liquid solution into the incineration chamber. Thus, the pecularities of Zählost formulations gave rise to the special design of the first incineration plant: a discontinuous double-chamber process.

4.1.2 Arsenicals

Special problems in adequate disposal of arsenicals are related to the toxicity of many arsenic compounds. Inasmuch as any type of incineration of arsenicals forms inorganic arsenic compounds of relatively high toxicity, a post-treatment step has to follow, before final disposal. Of the arsenical CW agents listed in Table 1, Clark I & II and Adamsite were used straight but also in solid mixtures with nitrocellulose. Other arsenicals, like arsine oil, are mixtures of several organoarsenic compounds; they sometimes were added to 'mustard' to function as anti-freeze (for winter use). Adamsite was not only the filling of shells and bombs; it has also been found as loose material in Munster in quantities of more than 40 tons. There are still more remainders in old salt mines or are expected to be found in our country. Like Zählost, the Adamsite has extremely low solubility in common, commercially available solvents. For this reason neither dosage as a liquid solution into the incineration chamber nor homogeneous reaction of Adamsite with any chemical reagent can be realized. Adamsite has been found more frequently than Clark I or II as World War II munition filling. Moreover, it is a common contaminant of the ground at former ammuntion production sites. There are some places near Munster with "hot spots" of more than one gram of total arsenic per kilogram of soil surface material. Because of the enormous chemical stability of Adamsite against hydrolytic or other degradation effects, the amount of inorganic arsenic compounds in the ground after more than 50 years is relatively small, about 0.1 - 0.4 ppm in groundwater. Possibly what inorganic arsenic concentration there is in the ground, and in groundwater, originated not in Adamsite degradation but rather from smouldering activities with arsenicals, during development of military field tests. Smoulder pots generate a considerable amount of products of arsenic oxidation, such as water soluble arsenic trioxide.

The extensive contamination of soil by organic arsenicals constitutes a serious problem of soil remediation. For this purpose and for destruction of solid arsenical agents, a second incineration plant is now under construction in Munster. Its function will be described in section 4.3.

4.2 TECHNOLOGY OF THE FIRST CW AGENT INCINERATION PLANT IN MUNSTER

4.2.1 Reasons for Plant Configuration

Taking into account the type and quantity of CW materials, we chose a two-stage batch-type furnace for their incineration. Primarily, this configuration was selected to solve the Zählost problem.

Canisters of Zählost that have been in the ground for more than 50 years have suffered almost complete destruction of the metal walls by corrosion. The surviving lumps of Zählost have "sulphur mustard" concentrations from as much as 25 % to as little as 0.1%, proceeding from inside to outside.

As mentioned, Zählost contains "sulphur mustard" analogs and various arsenic compounds, such as phenyl dichloroarsine (trivially referred to as "Pfiffikus").

These facts led to a process design with two special characteristics: a preliminary evaporation step to volatilize inherently volatile CW agents and other components, and a final arsenic precipitation process. These steps are followed by flue gas scrubbing.

The Munster incineration plant, after approximately five years of planning and construction, began full operation in 1980.

4.2.2 Preparations before Incineration

The isolated chemical agents of highly viscous consistency are placed, together with the empty metal casings, in 30 l polyethylene barrels and are, first of all, kept in an intermediate storage area designed to hold a week's supply. Contaminated waste is treated in the same way.

Solid chemical agents (e.g. Clark) removed from the opened shells are likewise delivered in plastic barrels, although only small amounts of them can be treated in the first incineration plant.

Prior to incineration, the materials are analyzed to determine their content of arsenicals. Based on the analytical results, batches are put together so as to optimize conditions for the most effective waste disposal and effluent air purification. The procedural flow within the incineration plant is depicted in Figure 19.

As the first step the polyethylene barrels are placed on a charging waggon (useful surface 2.9 x 0.9 m) with fire-proof bricks as insulation layer. Once the lids have been removed from the barrels, a waggon is moved into the first furnace chamber (evaporation chamber), via an airlock, by a pull-cable mechanism. Small amounts of liquid CW agents or liquid waste are put on the waggon in tin-plate boxes.

A negative pressure of 0.5 - 1.0 mbar, relevant to the atmosphere, prevails throughout the charging area, so that contaminated air cannot escape.

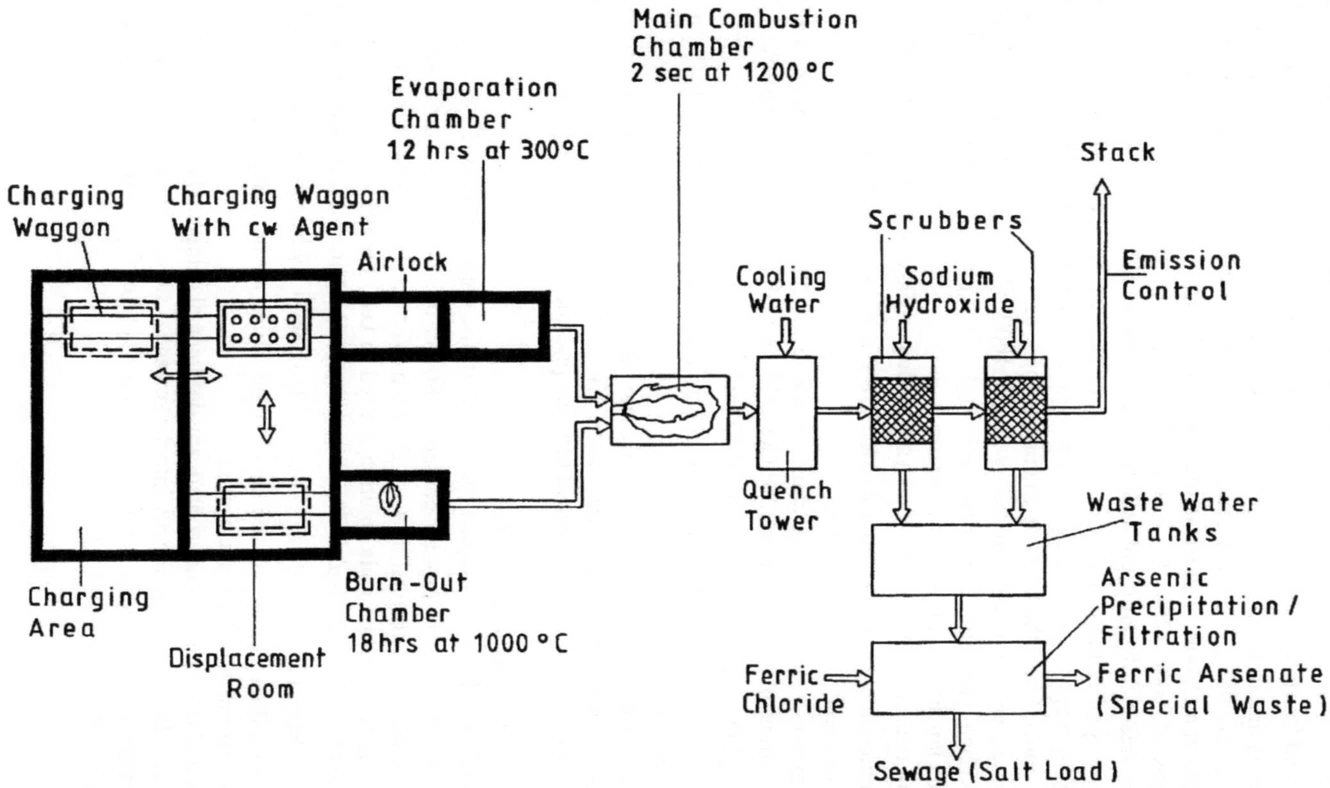

Figure:19 FUNCTIONAL DIAGRAM OF INCINERATION PLANT

4.2.3 Evaporation Followed by Combustion

Inside the evaporation chamber, the chemical agents are evaporated and partially subjected to pyrolysis at 300 °C in an inert gas environment (nitrogen plus CO_2 and H_2O produced by stoichiometric combustion of fuel oil). In the case of Zählost, this takes 14 hours due to the slowness of heat transfer and to prevent a sudden and violent combustion. By the suction of a central fan, the evaporated matter is moved into the main combustion chamber, which is lined with firebrick. It is completely burned at temperatures of 1000-1200 °C. A residence time of two seconds in the high-temperature area is fully sufficient to achieve total oxidation.

The main combustion chamber is heated with heating oil via an adjustable burner. The charging waggon containing the unevaporated organic-chemical and metallic residues is moved into the burnout-chamber, where especially metal parts, including cartridge cases, are completely eroded by heat at 1000 °C. They subsequently can be scrapped without presenting any chemical hazard. This process requires time about as long or even longer than in the first chamber.

"Sulphur mustard" (HD) is burned to generate carbon dioxide (CO_2) and water besides the environmental pollutants sulphur dioxide (SO_2) and hydrogen chloride (HCl), which have to be removed from the flue gas.

Calculation indicated that the combustion of Z-OA Zählost (see Table 6) gives a flue gas composition of: N_2, 71.40 mol.%; CO_2, 14.13 mol.%; H_2O, 8.46 mol.%; HCl, 4.27 mol.%; SO_2, 1.29 mol.%; As_2O_3, 0.45 mol.%.

The waste gas from the burnout-chamber is also processed through the main combustion chamber.

Fluid chemical agents can be fed directly into the main combustion chamber together with the fuel oil via a multi-fuel burner lance. In the process, the fluid is pumped from a charging tank through a double-walled heatable pipeline by a gear-type pump.

4.2.4 Flue Gas Cooling and Scrubbing

SO_2 and HCl are the most frequent pollutants in the combustion flue gas. To take them out, the flue gas is rapidly cooled to 80 °C by injection of cold water and subsequently washed with alkaline water in two packed scrubbing towers in series (each tower 6 m high). Caustic soda solution is used to keep the pH in the scrubbers at a constant level. The SO_2 and HCl are both neutralized in this way and the sodium sulfite from SO_2 is oxidized in a reaction tank with hydrogen peroxide (H_2O_2). Resulting are sodium sulfate and sodium chloride. Both are non-toxic, but they cause a salt load on the municipal sewage system.

If the initial material does not contain any detrimental compounds, the water used for scrubbing may be discharged directly into the sewage system. Unfortunately many CW agent formulations produced in former Germany are not free from arsenic compounds and need special treatment (see below).

In the meantime, the flue gas, after passage through an aerosol separator which retains any small droplets that have been carried over, is released into the open air through the stack.

Rates of emission of SO_2, HCl, NO_x (nitrogen oxides) and hydrocarbons are measured at the stack and recorded continuously by a computerized data aquisition system in the central operation control room. Dust particles, which may contain residual arsenic, for example, are separated isokinetically from the off-gas by a dust probe and subjected to laboratory analysis by atomic absorption spectroscopy.

The above-mentioned rapid cooling of the hot flue gas (from 1000° down to 80 °C) by the introduction of cold water is the reason that there is no recombination of molecule fragments to dioxins or difuranes. The so-called "dioxin window" (400-600 °C) is passed much faster than dioxins can be formed.

4.2.5 Arsenic Precipitation

If the material to be burned contains arsenic, as do some Zählost batches, in particular, the waste water must be subjected to a supplementary treatment.

Incineration of organic arsenic compounds produces arsenic trioxide, which is captured as trivalent arsenite in an alkaline environment. Such inorganic arsenic compounds are strong acute poisons with also carcinogenic potential. Conversion of arsenic into a compound insoluble in water requires oxidation to pentavalent arsenate, which can be effected by potassium permanganate (or hydrogen peroxide). Finally, after iron(III) chloride ($FeCl_3$) has been added, precipation of insoluble iron arsenate follows. The sludge-like material is dehydrated in a filter press and then filled into 200 l. barrels for final deposit in a special underground waste dump (a former salt mine).

Eventually the purified filtration waters are released into the municipal sewer system. However, because of their fairly high salt concentration, a waste water evaporator will be required in the future.

4.2.6 Operational Data

The incineration plant is operated 14 hours per day in two overlapping 8 hour shifts. During weekends and on holidays, the plant is kept warm, at a defined depressed temperature level, because starting up and cooling down the facility are time- and energy-consuming.

In the past few years, average availability ranged from 200 to 250 operation days a year; the plant was inoperative on about 50 days.

The throughput of the incineration plant varies considerably depending on the material incinerated. For "sulphur mustard" it is between 20 and 35 kg/h.

The energy demand for the combustion chambers amounts to 100 l per hour of fuel oil; also, 4500 kWh of electric power per day are required.

In case of emergency, the incineration plant has available a 258 kVA standby power generator and, for further safety, an accumulator station with a capacity of 700 Ah.

Inert gas for the evaporation chamber is produced by stoichiometric combustion of fuel oil and amounts to 400 Nm^3/h. [N stands for normal conditons: 0 °C, 1 bar] 60 Nm^3/h of compressed air at a pressure of 6 bar are used to 'atomize' the fuel oil.

Tap water (approx. 2 m^3/h) is used for flue gas cooling and the preparation of scrubbing solution.

The effective overall thermal power of the incineration plant is 1.2 MW.

4.2.7 Environmental Aspects

The CW agent incineration plant is subjected to continuous emission monitoring with attention to SO_2, HCl, residual hydrocarbons (including traces of chemical agent) and nitrogen oxides. Dust is collected at the stack by discontinuous sampling and analyzed for its arsenic content.

As far as the sewage is concerned, the pH of the discharged waste water is adjusted to approximately 8, and the arsenic content to less than 1 mg/l. The latter is the authorized threshold limit specified in the sewage regulations of the local communities. Some actual emission data of the incineration plant are summarized in Table 7. A comparison of the actual and the legally allowed levels of emission shows that the current chemical agent incineration process produces little pollution.

TABLE 7. Emissions from the first incineration plant in Munster

1. Flue gas		
Flue gas quantity	2800 m^3/h	
Flue gas temperature	105 °C	
Height of exhaust stack	30 m	
	Emission rates	**Limits (TA Luft 1986)**
HD	not detected	0.05 mg/m^3 (US TLV)
SO_2	< 5 mg/m^3	100 mg/m^3
HCl	< 5 mg/m^3	50 mg/m^3
As_2O_3	approx. 0.2 mg/m^3	1 mg/m^3
Dust	< 1 mg/m^3	30 mg/m^3
NOx	approx. 200 mg/m^3	500 mg/m^3
Dioxins	not detected	< 0.1 ng/m^3
2. Waste water		
Quantity	approx.	0.5 m^3/h
Salinity		5 - 40 g/l
pH value	adjusted to approx.	8
Arsenic	adjusted to	< 1 mg/l (limit)

4.3 TECHNOLOGY OF THE SECOND CW AGENT INCINERATION PLANT IN MUNSTER

4.3.1 General

Two major problems related to solid fillings of some chemical weapons established the need for another CW agent disposal plant, now under construction in Munster:

(1) solid chemical agents, particularly arsenicals, cannot be disposed of at sufficient throughput rates in the first incinerator, and

(2) the large concentrations of organic arsenicals and degradation products found in soils at many sites require a suitable soil cleaning process, for which purpose "laundering" followed by thermal destruction was chosen.

The chief shortcoming of the first incinerator with respect to the combustion of solid agents is its lack of poking devices. Without stirring of the solid material, there is only a little combustion in the surface layer; most of the material to be burned stays completely unaffected. That is a limitation so severe that, if only the first incinerator were used, it would take decades to destroy all of Germany's old CW agents. Furthermore, the arsenic concentrations found in soil samples from the former CW agent production sites, filling plants and test ranges are becoming recognized as a major environmental problem. Therefore it was deemed necessary by German environmental authorities to increase disposal capacity. Munster was chosen as location for the second plant because the major disposal problems are in its vicinity and also because the plant is considered to be a technical supplement of the first one.

Only hazard assessment programmes related to contaminated areas like the Military Training Ground Munster North, where the special heritage of the two World Wars is located, will determine whether remedial actions are required or not. Unfortunately most of the cases discussed, sometimes emotionally, in public still lack a precise hazard assessment. Quantitative data on sites contaminated with chemical agents, e.g. arsenicals and their decomposition products, are only fragmentary, mostly concerning surface soil samples that were taken at "hot spots". Such places often are evident to the observer by their absence or near-absence of vegetation. Because of buried munitions and heavy contamination in places that contain one to ten grams of total arsenic per kg of dry soil, the so-called "Red Areas" have been surrounded by wooden fences, but not examined in detail, for instance in regard to depth of contamination. One very important indicator, the arsenic concentration of the ground water, is however determined and recorded regularly by sampling groundwater from a large number of observation wells.

Inasmuch as no alarming increase of the arsenic concentration has been noticed, the ecological influence of the arsenical CW agents must be characterized as an extremly long-term one. Moreover, little is known about degradation and transport mechanisms in the Munster area. Nevertheless, the high concentrations of arsenic in the ground must be judged to be a serious long-term environmental problem. Besides the groundwater pollution hazard, arsenic dust may be dispersed by the wind. Another influential aspect is what future use may be made of the terrain. For some intended uses, remediation of sites may be urgent.

4.3.2 Soil Washing and Flotation

A novel approach to remediation of contaminated soil, particularly from Munster North, will involve, in sequence a soil washing and flotation step, to eliminate contaminants from the dredged soil, and then a thermal destruction and vitrification step using high temperature plasma technology. The plant is expected to be ready for processing in 1998.

The objective of soil washing is to separate contaminants and harmful substances from the soil particles and to remove them together with the silt, which is composed of the smallest-particle fractions. Three factors must be combined for good separation results:

(1) water acting as transport medium and as solvent for soluble inorganics,

(2) application of mechanical energy to dislodge contaminants from soil grain surfaces, and

(3) use of auxiliary agents: flotation additives (collectors and foamers) to make the contaminants skimmable, and complexing agents to sequester heavy metals as water-soluble compounds.

Water itself is an ideal auxiliary reagent for the elimination of contaminants, because it is a natural soil constituent and residual water need not be removed after soil washing. Contaminants vary in their mode of association with soil particles. There may be agglomerates/soil inclusions, which require the shearing forces of a strong water jet to disintegrate them. This results in a suspension of both contaminants and soil particles. The contaminants may also be incrustations on soil grain surfaces. This is the form in which arsenical CW agents are most likely attached to the particles at the soil surface, particularly when the agents were burned on the ground at low temperatures, as in smoulder-pots. In a soil washing process frictional stress, supported by surfactants - as effected in wet disintegration systems - finally leads to detachment of the contaminant incrustations. The efficiency of this process decreases with decreasing particle size of the supporting soil material.

Liquid or greasy substances (oil films) can also be separated from the soil grains by wet disintegration. Resulting is emulsive waste water. Machinery that disintegrates aggregates of soil grains with contaminants by shearing, friction, or impact stress is especially suitable for soil washing and will be integrated into the second plant in Munster. The operational suitability of such devices depends, of course, on the respective coarse and fine structures of the soil matrix and on the physicochemical character of the soil contamination. The selection of appropriate auxiliary agents also requires extensive technical experience and repeated laboratory tests.

In separation by flotation, the washed-out and suspended contaminant particles (or droplets) adhere to air bubbles and accumulate in a foam layer at the surface of the washing water. The foam is removed with a skimmer. After defoaming, it is combined with the silt to constitute the contaminant concentrate that is to be treated in the second incineration plant.

The soil washing process involves a number of conveyor, classifying, washing, settling and dehydration units. The soil of former CW agent disposal sites is, as a rule, not a homogeneous solid material but contains stones, vegetation material (stubs), lumps of concrete, metal residues and ammunition fragments. It is obvious that before entering such a plant, such soil must be sorted with use of detection devices (e.g. metal detectors), for both procedural and safety reasons. Considerations of economy and environmental protection require that the washing water be recycled. Water that is highly contaminated with dissolved soil constituents and harmful substances like arsenites undergoes water treatment.

For soil treatment at the Munster North Training Area, a cleaning procedure based on multiple flotation stages will be used (Figure 20). Tests conducted with a pilot plant showed that most of the arsenic contamination could be concentrated in a fraction constituting 10 to 15 % of the soil treated. Only this contaminated fraction would then have to be processed thermally. Otherwise it would be economically impossible to restore the enormous volumes of contaminated soil on the Munster North Training Area or on other former CW agent disposal sites.

4.3.3 Thermal Treatment of Contaminated Soil

Thermal procedures for soil treatment include: (1) fluidized-bed furnaces; (2) chain-grate furnaces; (3) direct-fired drum furnaces (rotary kilns) with afterburners; (4) indirect-fired drum furnaces (pyrolysis furnaces) with afterburners; and (5) plasma high-temperature technology.

Except for the last procedure, these have been tested many times. Most of the thermal treatment plants - which are mainly designed as central facilities and used in The Netherlands and the USA - are based either on the direct or indirect procedure. In general, thermal plants are used to clean soil masses contaminated with the following substances: contaminants based on mineral oil and pit-coal secondary products; cresols; residues of synthetic dyes; polyaromatics; complexed cyanides; and contaminants based on halogentated (chlorinated) hydrocarbons, including PCB's, dibenzo-dioxines/dibenzofuranes, DDT and other insecticides.

Temperatures required for incineration of these substances range from 550 °C for contaminants based on mineral oil up to 1200 °C. If the contamination is physically inhomogeneous, and variable in composition, it is advisable to use high-temperature incineration. In contrast to hazardous waste incineration plants, thermal soil treatment plants require a 100 percent energy supply, since the heating value of the contaminated soil is practically zero. However, in thermal treatment of contaminated soil better control is possible due to a more balanced condition of the combustion process.

The thermal process to be installed in the second incineration plant in Munster for disposal of arsenic concentrates from soil washing and flotation will operate with a plasma burner. The same facility will also be used for destructive disposal of solid chemical agents like Adamsite, Clark I and II and chloroacetophenone.

The high-temperature plasma technology (PLASMOX® process) is used in metallurgy and plays an important part in the production of titanium alloys. The vacuum melting procedure used for compaction of radioactive waste is also based on plasma technology.

MGC-PLASMA Ltd. in Muttenz, Switzerland has developed (under licence from Californian Retech Inc.) a process technology that uses plasma burners to treat hazardous waste and contaminated soil in the available high-temperature zone. A special feature is that the residue is completely harmless and does not need to be disposed of at a hazardous waste dumping site. In this case arsenic will be firmly bound in a glass matrix as slag.

By means of electric current or in an electromagnetic field, the system is heated up until temperatures of 10,000 to 20,000 °C are reached, in an electric arc at the top of a plasma torch. Of the various plasma burners that have been developed, only the direct

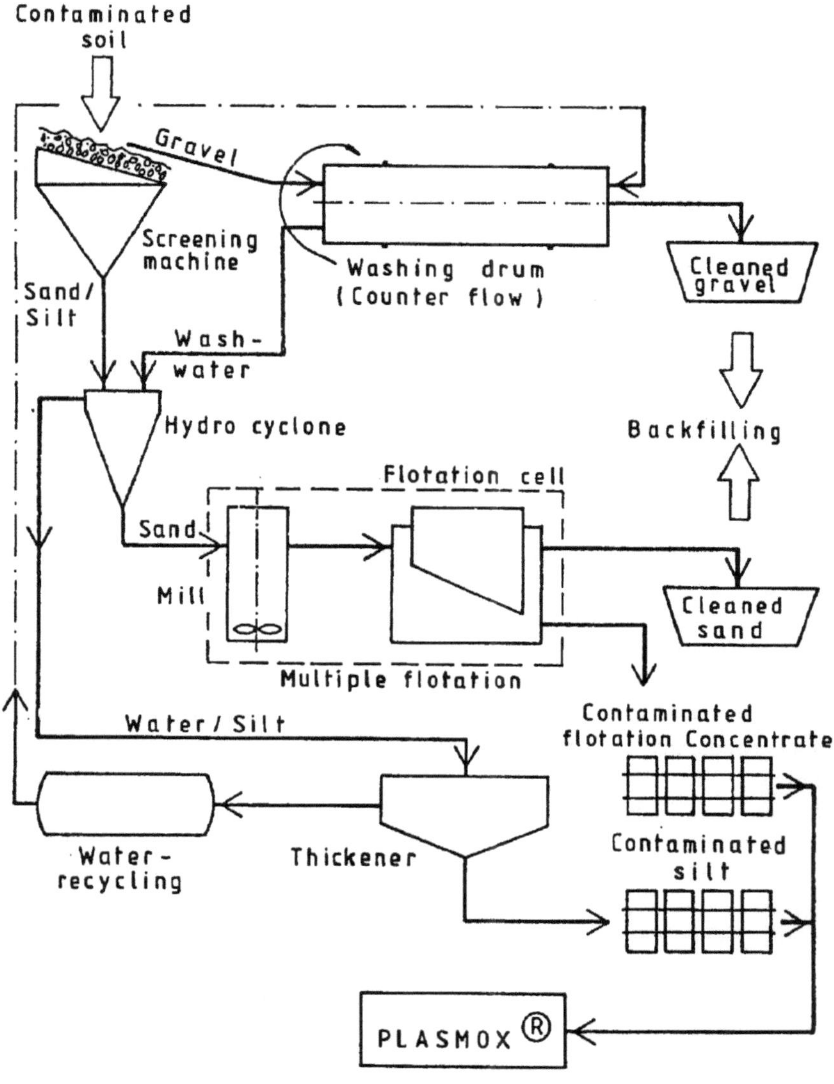

Figure: 20. Soil Cleaning Plant

burner version, in which the anode is the material to be heated in a rotating centrifuge, is satisfactory to achieve the desired melting of fed-in waste material. In an electric arc such as that of the PLASMOX® process, with high local energy density, the material is heated up to about 1600 to 1800 °C. These temperatures are higher than be attained in a direct-fired rotary kiln.

The plasma high-temperature reactor (Figure 21) consists of a cylindrical chamber with a centrifuge at bottom containing the fed-in material (approx. 4 t). The arc does not burn the rotating organic chemical material but causes it to pyrolyze. After 4 hours, the centrifuge is stopped, and the melted material is emptied through the central discharge opening into a slag cart. Heavy metals like arsenic remain in the slag where they are evenly distributed and firmly bound. As mentioned above, this slag is similar to glass. The various elements contained in it are released in only extremely small quantities into the leach test eluate. As an inert substance, the slag can be used as a filling material in civil engineering (e.g., gravel for road construction) and requires no expensive disposal at a hazardous waste dumping site. The above-mentioned leach test (EPA Standard Leach Test) is performed by treating the slag in a slightly acidic environment to simulate the time effect of acid rain on the dumped material.

Feeding of the concentrates with the arsenic contaminants and of solid chemical agents, by using a screw feeding pump or a cellular wheel lock, is designed to run continuously. Experimental runs will show whether the solid agents can be introduced by addition to the regular material flow or have to be treated separately. It is also possible to mix them with less contaminated soil in order to avoid problems at the flue gas scrubbing step that might arise if the arsenic load were too high.

The flue gases produced during the pyrolysis process pass through the central opening and enter the oxidation chamber in which oxygen is injected to burn residual organic gases. This oxidation chamber of the plant is also heated by a plasma burner. Subsequently the flue gas is passed through a cleaning unit and freed of nitrogen oxides in a DENOX facility. The flue gas volume is relatively small since it does not include gases arising from the combustion of fossils fuels; they are not required for this facility.

Furnace systems must have effective flue gas cleaning units in order to eliminate harmful substances such as SO_2, HCl, HF, NO_x, arsenic trioxide and dust produced during incineration. Together with the waste heat utilization units, these plant components are often of a much larger size than the incineration unit itself.

In view of the very satisfactory function of the aqueous flue gas scrubbing system of the first incineration plant, it was decided to equip the second disposal line also with scrubbers. As a novelty, the collected scrubbing waters will be evaporated to dryness. Hence, the second incineration plant will produce no waste waters. Eventually the salt residues will have to be deposited as toxic waste.

The design of the second incineration plant as a whole is sketched in Figure 22. It becomes obvious that the soil washing process is advantageous with respect to both cost and consequences. Only 10%-20% of the incoming quantity of contaminated soil has to be treated further in the high-temperature plasma furnace. This is more economical than to heat up all of the contaminated soil. As an ecological advantage, the 80 to 90% of laundered soil can be returned directly to the ambient terrain.

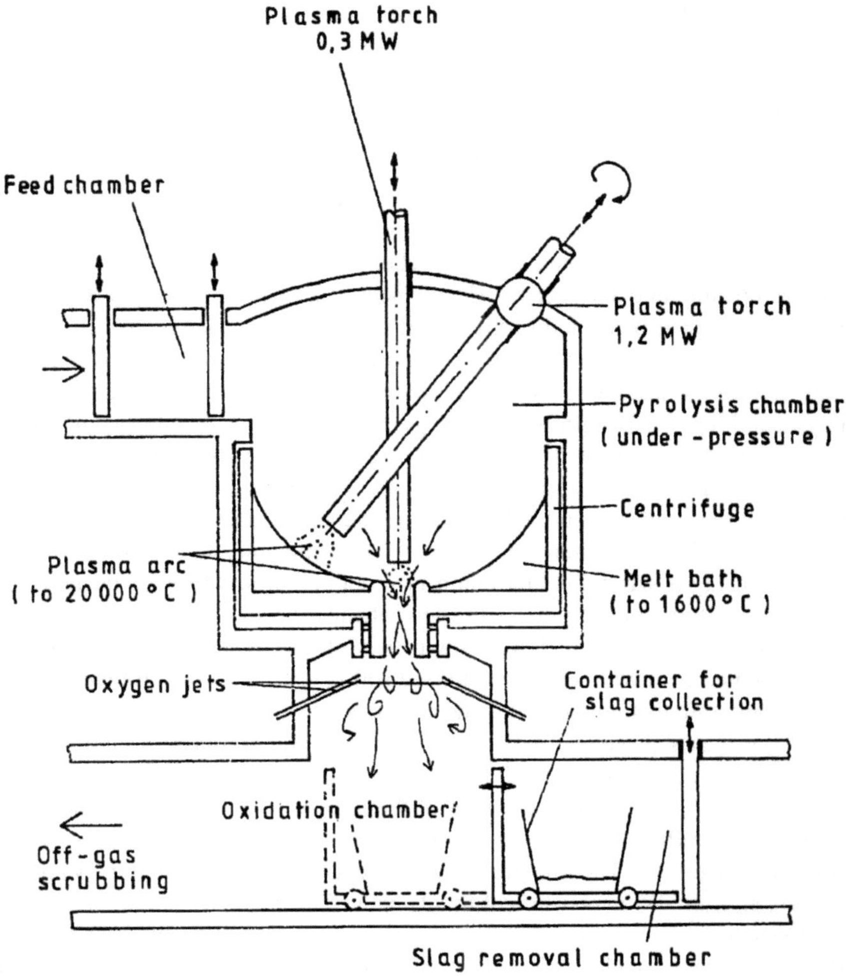

Figure 21: Schematic Plasmox® Reactor

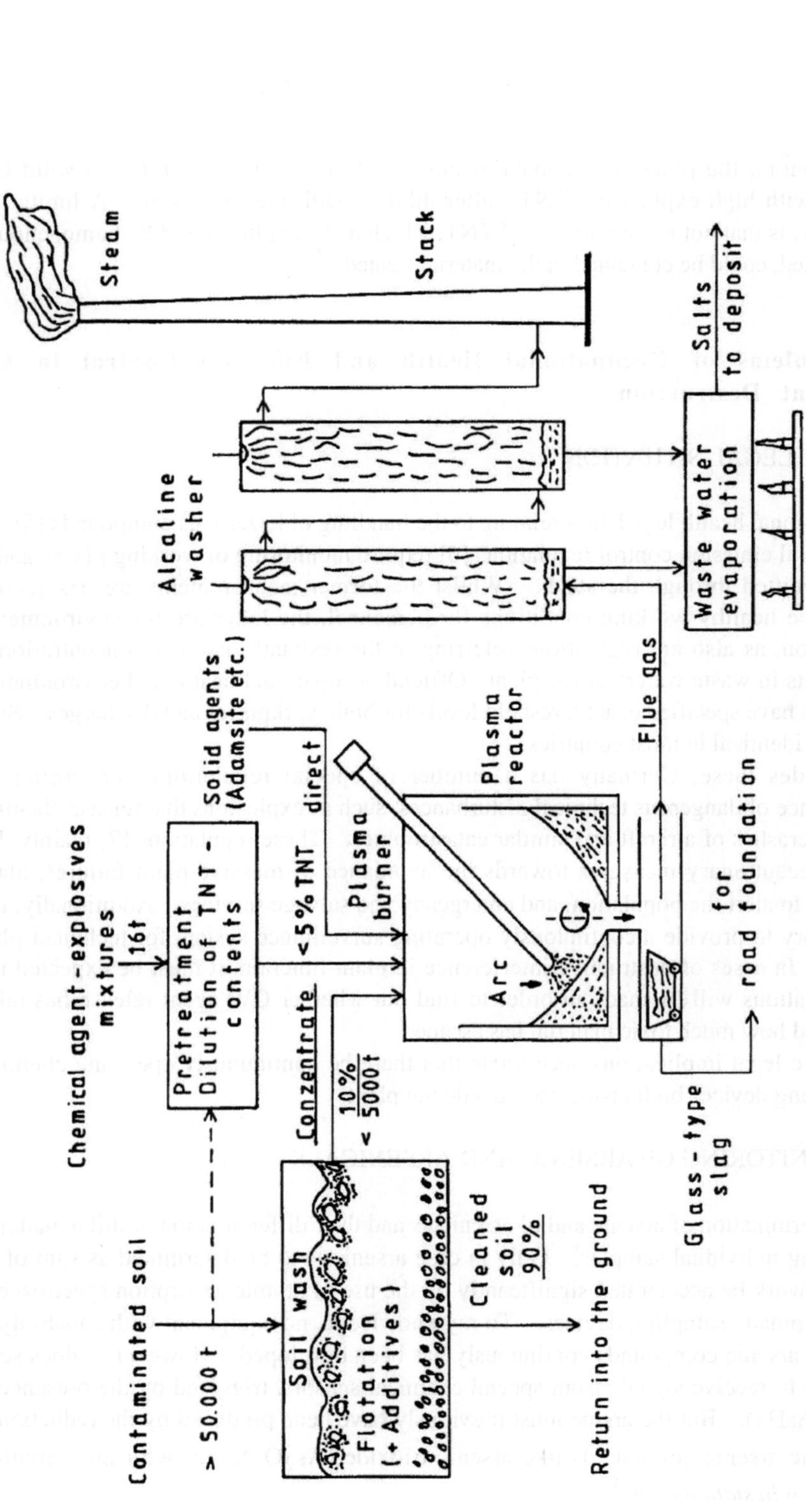

Figure: 22 Schematic Plan of the 2nd Incineration Plant

Furthermore, the process will make it possible to dispose of mixtures of solid CW agents with high explosives (TNT), after dilution with inert material. A limitation, however, is that not more than 5% of TNT, which additionally should be homogenously distributed, could be contained in the material treated.

5. Problems of Occupational Health and Emission Control in CW Agent Destruction

5.1 THE LEGAL SITUATION

Occupational health legislation relating to the handling of hazardous compounds [5] and the federal emission control regulations [6] require monitoring of working places and of gases emitted through the stack. Whilst the former measurements are designed to guarantee healthy working conditions for personnel, the latter are for environmental protection, as also are regulations referring to the residual load and concentration of pollutants in waste waters of the plant. Official occupational health and environmental agencies have specified exact threshold levels for both workplaces and discharges. Such data are identical in most countries.

Besides these, Germany has a number of special regulations concerning the occurrence of dangerous technical disturbances, such as explosions that release chemical agents, crashes of aircraft and similar catastrophies. These regulations [7] mainly deal with precautionary measures towards the avoidance of massive plant failures, alarm systems to alert the population, and emergency and salvage services. Additionally, it is obligatory to provide a continuously operating surveillance system for technical plant safety. In cases of destructive interference in plant function, it must be expected that investigations will be made in order to find out whether CW agent release has taken place and how much toxic material has escaped.

These legal implications necessitate that there be continuously operating chemical monitoring devices both inside and outside the plant.

5.2 MONITORING OF ARSENIC AND ARSENICALS

The determination of arsenic and of arsenicals and their differentiation is still a matter of analyzing individual samples. Only in case arsenic is to be determined as sum of As can the work be accelerated significantly by the use of atomic absorption spectroscopy with automatic sampling devices. To my knowledge, no equipment with capability to monitor arsenic compounds continuously has been developed. However, it does seem possible to receive signals from special chemical sensors, triggered by the presence of arsine (AsH_3). But the arsine must previously have been produced by the reduction of inorganic arsenic compounds like arsenic trioxide (As_2O_3), e.g. with zinc granules (hydrogen *in statu nascendi*).

Organic arsenicals, which are mostly solid materials, must undergo shattering of their molecular structures before they can be determined as inorganic compounds.

Obviously, shattering reactions cannot easily be run as continuous processes. For this reason, dust samples are collected from the flue gas inside the stack using time-programmable air sampling equipment. The same would be possible by establishing monitoring circles around plants in which arsenicals are processed. The radii of such circles should be such that, when the disposal plants are operating normally, atmospheric concentrations of arsenic compounds will approximate the maxima for human exposure permitted by environmental regulations. At present, this can be realized only by means of remotely positioned air samplers. That is a reasonable solution for long-term monitoring, as well as for quantitative registration of a sudden break-out of toxic vapours, dust or aerosols. This solution depends on the assumption that there is no appreciable long-term background of the respective pollutants. Nevertheless, interchanging the sampling tubes and paper discs followed by chemical analysis makes a high work-load for personnel. For this reason new technical principles of on-line, real-time plant surveillance should be developed, to enable automatic triggering of safety measures, in the case of an unnoticed escape of toxic effluents.

5.3 MONITORING OF "MUSTARD GAS"

"Mustard gas" [bis(2-chloroethyl) sulfide] is more amenable to continuous monitoring than are arsenicals. At the CW destruction plant in Munster, the absence (or unexpected occurrence) of 'mustard' in the flue gas of the stack is recognized by an automatic flame ionization detector. For breakthrough surveillance of the charcoal filter elements of the CW munition dismantling building in Munster, we have recently changed to RAID 1 ion mobility dectectors (from Bruker-Saxonia, Leipzig, Gerrnany), which operate more reliably than flame photometric detectors. They avoid the frequent false alarms that can be caused by small molecule, volatile sulphur compounds. Moreover, RAID I permits the qualified differentiation of breakthroughs arising from "sulphur mustard" (HD), cyanogen chloride (CK), phosgene (CG), hydrocyanic acid (AC) or chloropicrin (PS).

For the same purpose and for monitoring stack emissions is another hardware product, the MINICAMS® automated monitoring and alarm system (from CMS Research Corp., Birmingham, Alabama, USA). This system was developed specifically for CW agents and simulants like GA, GB, GD, HD, VX, DFP, DIMP and DMMP. It is based on sampling of vapours of the above-mentioned compounds by collection on solid adsorbents; the desorbed compounds are separated by capillary gas chromatography with flame-photometric or flame-ionization detection. MINCAMS® can detect as little as 0.01 ng of nerve agent.

Finally it should be mentioned that multipoint ambient air monitors and multicomponent analysis systems are appearing on the analytical instrumentation market. Such systems mostly are based on Fourier transform infrared (FT-IR) spectroscopy or are on-line systems with mass spectrometers. Although such systems enable up to 20 components to be analysed simultaneously, their usefulness for plant surveillance is questionable because of the cost of establishing numerous sampling points around toxic agent disposal facilities.

6. Summary and Conclusion

In Germany a well-organized, well-structurized technical system exists for the disposal of CW-related remnants from two World Wars. The installations involved have been centralized in the Munster region in the north of the country. The munition items can be manipulated safely in a dismantling facility in order to separate the CW agents from the other fillings for further destruction. A CW agent incineration plant has been in function since 1980. It serves for the thermal combustion of liquid (and viscous) CW agents as well as for the disposal of different types of waste from dismantling and decontamination activities.

A second facility for incineration, preceded by soil washing, is under construction. This plant, which will be ready about 1998, is designed for the disposal of solid CW agents and mixtures of CW agents with explosives. Its main task, however, will be the remediation of soil with warfare-related contaminants, primarily arsenicals and their degradation products.

The processing of old chemical munitions and their fillings is characterized by a large diversity of small but all the more important technical steps. Improvements are possible and desired in a diversity of fields, such as automation of mechanical operations, improvement of combustion technology, and development of self-operating surveillance systems. Whilst the actual quantity, in tons, of CW agent to be destroyed is relatively small, an unusually high safety standard is required. That is why CW agent disposal is rather expensive in comparison to other areas of chemical engineering.

7. References

1. Lohs, K. and Martinetz, D. (1995) Karnpfstoffmunition, Fachinformation Rüstungsaltlasten II - 2.6, *Rippen-Handbuch Umweltchemikalien* - 32. Erg.Lfg. 12/95, Ecomed Verlagsgesellschaft, Landsberg
2. Franke, S. et al (1994) *Chemie der Kampfstoffe Umwandlung und Vernichtung.* Gesellschaft für Kampfmittelbeseitigung Dr. Koehler mbH, Munster/Burg
3. Mills, A.K. (1945) Investigation of Chemical Warfare Installations in the Munsterlager Area, including Raubkammer. Combined intelligence objectives sub-committee G-2 division, SHAEF (Rear) APO 413 CIOS item 8, Chemical Warfare/UK
4. Baltic Marine Environment Protection Commission, Combatting Committee, 11th meeting, Copenhagen, Denmark; 13-16 October 1987
5. Verordnung zum Schutz vor gefährlichen Stoffen (Gefahrstoffverordnung - GefStoffV), issued 26 October 1993.
6. Gesetz zum Schutze vor schädlichen Umwelteinwirkungen durch Luftverunreinigungen, Geräusche, Erschütterungen und ähnliche Vorgänge (Bundes-Immissionsschutzgesetz-BImSchG), issued 15 March 1974
7. Zwölfte Verordnung zur Durchführung des Bundes-Immissionsschutzgesetzes (Störfall-Verordnung) - 12.BImSchV), issued 27 June 1980

PRACTICAL ACTIONS OF RUSSIA ON PREPARATIONS FOR DESTRUCTION OF STOCKPILED LEWISITE AND 'MUSTARD'

GENERAL-COLONEL S. V. PETROV, Candidate of Technical Sciences;
GENERAL-MAJOR V. I. KHOLSTOV, Professor Doctor of Chemical Sciences;
COLONEL V. P. ZOUBRILIN Professor Doctor of Technical Sciences;
N. V. ZAVIALOVA Doctor of Biological Sciences
Ministry Of Defence Of Russian Federation

In the periods of World Wars I and II and the Cold War, many developed states, including the former USSR, engaged in the mass production of chemical weapons. The resulting vast stockpiles of chemical weapons (CW), as years passed, induced serious anxiety in the world community. Accordingly many states that were the owners of those deadly stockpiles made internal and international commitments to destroy their chemical weapons (CW) as soon as possible.

On January 13, 1993, Russia and other progressive states of the world community signed the Convention on the Prohibition of the Development, Production, Stockpiling and Use of Chemical Weapons and Their Destruction (CWC). By signing this document, Russia once again confirmed its adherence to the principle of CW disarmament and demonstrated good will, striving for general and complete CW destruction in the near future.

As events have shown, the destruction of CW stockpiles is an extremely complicated mission in both its social and technical aspects. It is very important to the world community that Russia carry out provisions of the Convention on CW destruction.

Making use of bilateral and multilateral international contacts, as well as its own experience, the Russian Federation has persistently searched for acceptable ways of solving the remarkably complicated problems in this sphere.

The Russian Federation's international obligations in the sphere of CW destruction presuppose that we have the required financial and material resources, the proper production base, infrastructure, and specialists to carry out safely the technological processes of chemical agent detoxification. Taking into account that Russia is responsible to the world community for CW destruction and that the problems of CW

J.F. Bunnett and M. Mikołajczyk (eds.), Arsenic and Old Mustard: Chemical Problems in the Destruction of Old Arsenical and 'Mustard' Munitions, 79–90.

destruction are of great importance for our country, on March 24, 1995 Russian President Yeltsin issued an edict "On the Preparation of Russian Federation for Dischargement of International Obligations in the Sphere of Chemical Disarmament". The edict implied organization of works for CW destruction. Accordingly, "A Federal Program of RF CW Stockpiles Destruction" was worked out. It is planned that CW stockpiles be destroyed in specially created facilities in CW storage areas.

This program, an expression of national concerns as expressed in political discussions, serves to reduce social tensions and to protect the environment. The presence of vast CW stockpiles is undesirable in many respects.

Thus, preparing to meet its international obligations while also pursuing national policy, Russia has given priority to destruction of its own CW stockpiles as a major national objective. Preparation for CW destruction in Russia started already in the middle 1970's. The first tangible result in this program was creation of a CW destruction mobile complex. In the late 1980's, using the technology of the complex, the first experimental CW destruction facility was built near Chapaevsk city. It was outfitted with state-of-the-art equipment of that period. However, for social and political reasons the facility was not put into work. It has been reconstructed into a training centre for working out CW destruction technologies and for preparation of specialists to operate such facilities.

Thus, in the early 1990's Russia had insufficient technical base to begin CW destruction. That is why, under current circumstances, we are proceeding deliberately in developing a broad program of CW destruction. We are addressing various problems stage by stage. Once good processes have been developed, facilities must be constructed and then put into operation.

Technical base development will have been carried out taking into account a number of factors relevant to creating destruction facilities.

Sites chosen for facilities construction should be submitted to proper authorities and to citizens for their approval.

An early stage involves creation of a technical base for destruction of old blister agents ('Mustard', Lewisite and 'Mustard'-Lewisite mixtures), which are the most dangerous to store.

Another consideration is that it is necessary first to destroy the stockpiled 'Mustard'-Lewisite mixtures stored in high tonnage containers. Two factors underlie this decision: (1) munitions with thick walls provide relatively much safer chemicals storage than the relatively less thick walled high tonnage containers; (2) In case of an unforeseen accident, the amount of chemical released from one high tonnage container will exceed by 100 to 1000 times the amount released from a munition.

Next in priority is design and construction of works for organophosphorus agent destruction. This will however be carried out simultaneously with work on blister agent destruction; it again requires choosing sites and their submission to proper authorities and to citizens for their approval. In choice of technologies and site for CW destruction, safety is of paramount importance. Major attention is paid to the safety of citizens and to preservation of the environment.

1. Russian Chemical Weapons Stockpiles

Russian CW stockpiles include aviation, rocket artillery and artillery munitions equipped with chemical agents, as well as chemical agents stored in bulk containers. Altogether, the chemical agents in the CW weigh about 40,000 tonnes. According to CW types, the division is as follows: 11.7 thousand tonnes of Sarin, 4.8 thousand tonnes of Soman, 15.2 thousand tonnes of VX, and about 7.7 thousand tonnes of Lewisite, 'Mustard' and 'Mustard'-Lewisite mixtures.

Information concerning the different types of Russian chemical munitions, and their classification, was presented to delegates from 45 states during a 1987 visit at the Shikhany Military Establishment.

The CW agents in the stockpiles are classified according to storage mode: VX, Sarin, Soman, and Phosgene: 100% in munitions and devices; 'Mustard': 100% in containers; Lewisite: 2% in munitions and devices and 98% in containers; 'Mustard'-Lewisite mixtures: 40% in munitions and devices and 60% in containers.

This distribution of CW agents in Russian CW stockpiles is, according to location in one of seven depots in different areas, is: in Kambarka, 15.9% (of Lewisite); in Gorny, 2.9% (of 'Mustard', Lewisite, and 'Mustard'-Lewisite; in Khizner), 14.2% (of VX, Sarin, Soman, and Lewisite); in Maradykovsky, 17.4% (of VX, Sarin, Soman, and 'Mustard'-Lewisite); in Pochep, 18.8% (of VX, Sarin, and Soman); in Leonidovka,17.2% (of VX, Sarin, and Soman); in Shchuchye, 13.6% (of VX, Sarin, Soman and Phosgene).

Lewisite and 'Mustard' stockpiles are stored at four depots: in Kambarka, in Khizner (Udmurt Republic), in Maraykovsky (Kirov region), and in Gorny (Saratov region).

Besides 'Mustard' and Lewisite, stockpiles of blister chemicals also include 'Mustard'-Lewisite mixtures. Moreover, the stored 'Mustard'-Lewisite mixtures have different structures; for example, in one case it's a dual component 'Mustard'-Lewisite system, while in another a mixture of 'Mustard'-Lewisite in the solvent dichloroethane.

Distribution by weight of blister chemicals located at the four depots earlier mentioned is as follows: Gorny, in high-tonnage containers 293 tonnes of Lewisite, 807 tonnes of 'Mustard', and 157/71 tonnes of 'Mustard'-Lewisite mixtures (the denominator shows the number of tonnes of 'Mustard'-Lewisite mixture in dichloroethane solvent); Kambarka, in high-tonnage containers 6349 tonnes of Lewisite; Khizner, in artillery munitions 129 tonnes of Lewisite; Maradykovsky, in aerial bombs 149 tonnes of 'Mustard'-Lewisite mixtures.

As was presented, most of the Lewisite and 'Mustard' is stored in high-tonnage bulk containers. About 280 tonnes of blister chemicals are stored in artillery munitions and aerial bombs.

In spite of the fact that storage of old blister chemicals causes more difficulties in respect to safety than storage of other chemicals, all these sites are absolutely protected from any technological accident. This situation is achieved by strict control over the condition of containers and munitions. Much attention is paid to high-tonnage containers. Each container has a control point, which helps periodic examination of the thickness of container walls. To monitor possible leakage of the chemical agents from containers, munitions and bombs, tests with a special indicator paper are frequently made.

Security arrangements at storage sites practically exclude any possibility of terrorist efforts to capture chemicals.

2. Technological Aspects Of Lewisite And 'Mustard' Destruction

Destruction of CW in our country causes anxiety among most of its citizens. They lack confidence in the competence and reliability of the authorities in respect to choosing processes and designing facilities to destroy CW agents. Thus, alternative technologies for Lewisite and 'Mustard' destruction (utilization) were proposed to executive power bodies and to communities near planned sites for CW destruction. Objections were raised, even though those technologies had been worked out by different Russian scientific research and design organizations and establishments, in competition, according to requirements of ecological and technological safety.

2.1 LEWISITE STOCKPILE DESTRUCTION TECHNOLOGIES

Several technological methods for Lewisite destruction have been proposed by different organizations of the Russian Federation.

2.1.1.*Treatment with Sulphur.*

This method is based on the interaction of Lewisite with sulphur at high temperature, which produces a tough polymeric mass, appropriate to bury in special burial grounds. The reaction goes according to the following scheme:

$$\mathbf{ClCH = CHAsCl_2 + S \longrightarrow Polymer + HCl}$$

Reaction temperature is 160-180°C, the Lewisite/sulphur mass ratio is 1/3, and reaction time is 30-40 minutes. The resulting reaction products, after coagulation, are cast in sulphuric alloy block form, which can be buried in special burial grounds.

The main disadvantages of this method are that it generates a vast amount of arsenic-contained waste requiring burial, and that the burial grounds must be isolated from aquifers beneath the surface.

2.1.2. *Lewisite Interaction with Ethylene Glycol Monomethacrylic Ester and Its Subsequent Polymerization.*

This reaction, in the presence of a hydrogen chloride acceptor, with subsequent radical polymerization of the initial products, leads to the production of a solid, water-insoluble polymeric mass.

$$ClCH = CHAsCl_2 + CH_2{=}\underset{\displaystyle CH_3}{\underset{|}{C}}-\overset{\displaystyle O}{\overset{||}{C}}-O-CH_2CH_2-OH \searrow \text{Polymer} + HCl$$

The process goes at low temperature (60-90°C), tributylamine can be used as hydrogen chloride acceptor, and azo-bis-isobutyronitrile can be used to initiate polymerization. As in case of the sulphur treatment, the polymer produced requires burial. This method thus has the same disadvantages as does the sulphur treatment.

2.1.3. *Chlorination.* Interaction of Lewisite with gaseous chlorine at 110-130°C causes production of arsenic trichloride and a mixture of chlorinated hydrocarbons.

$$ClCH = CHAsCl_2 + Cl_2 \longrightarrow ClCH{=}CHAsCl_4$$

$$Cl_2CHCHCl_2 \xleftarrow{Cl_2} ClCH = CHCl + AsCl_3$$

This method is very interesting because the main reaction product is arsenic trichloride, from which we have, in our country, worked out a technology to produce extra-pure arsenic and gallium arsenide. The main shortcomings of the method are: that conversion of Lewisite is not entirely complete, and difficulties in separation of the chlorinated hydrocarbons, which are poisons.

2.1.4. *Alkaline Hydrolysis.*

This method, which employs aqueous alkali, has potential to achieve full decomposition of Lewisite. Subsequent electrolysis of the products yields useful products. The acetylene produced is blown away by inert gas. Reaction occurs during an hour at temperature 103-110°C. The process scheme, including electrolysis stage, is the following:

$$2\ ClCH = CHAsCl_2 + 6\ NaOH \longrightarrow 2\ As + O_2 + 2\ HC{\equiv}CH + 2\ H_2 + 3\ NaOCl + 3\ NaCl + H_2O$$

An especially useful final product is arsenic. Conceptually the process of metallic arsenic production consists of two stages: Lewisite hydrolysis, and electrolysis of arsenious acid in aqueous sodium chloride solution. This method is reliable for Lewisite destruction, and has the virtue of producing arsenic, which is valuable, and suitable for longterm storage.

2.1.5. *Alkoholysis.*
As a result of reaction with alcohol solutions of sodium alkoxides, Lewisite is decomposed into acetylene, sodium chloride, and a trialkyl arsenite:

$$ClCH = CHAsCl_2 + 3\ RONa \longrightarrow (RO)_3As + HC \equiv CH + 3\ NaCl$$

The products appear in three different phases, which facilitates their separation. The reaction goes at 50-80°C, with Lewisite and alkoxide in mass proportion 1.0 : 3.6. Duration for full detoxification is 2-3 hours. After separation of the reaction products, the acetylene can be burnt. The trialkyl arsenite, after purification, can be used in production of pure arsenic, and sodium chloride after separation from arsenic compounds should be suitable for production of chlorine and hydrogen via electrolysis.

The method has value because it can effect complete Lewisite destruction and enables utilization of several reaction products. Nevertheless, this method should not be used for mass destruction of Lewisite, because the toxic and flammable trialkyl arsenite produced requires further treatment; it is not suitable for long-term storage.

2.1.6. *High Temperature Oxidation.*
This technology has been put into practice in the USA and in Germany, and in Russia it is an approved high temperature oxidation technology. It is a matter of common knowledge that Lewisite can be completely oxidized in the flame of a gas burner at high temperatures (exceed 600°C). The process goes as follows:

$$ClCH = CHAsCl_2 + C_3H_8 + 7.75\ O_2 \longrightarrow 0.5\ As_2O_3 + 5\ CO_2 + 2\ HCl + 0.5\ Cl_2 + 4\ H_2O$$

The method has value as a reliable method for Lewisite destruction, and for the possibility of recovering the arsenic oxide, which can be stored in the long term. However Lewisite, as an organic chlorine compound, can while burning produce toxic chemicals such as dioxins.

2.1.7. *Hydrogenolysis.*

This method involves destruction of Lewisite by treatment with hydrogen (diluted with nitrogen) at temperatures of 800-1000°C. In the course of reaction, metallic arsenic, arsenic trichloride, hydrogen chloride, carbon and hydrocarbon fractions may be produced. The product gases are treated so as to effect arsenic condensation, collection of arsenic trichloride, hydrogen chloride absorption, and separation of hydrogen from hydrocarbons by the method of membrane technology. Recovered hydrogen returns to the reactor. The hydrocarbons can serve for production of hydrogen by the method of steam conversion.

The reaction scheme may be as follows:

$$ClCH = CHAsCl_2 + H_2 \xrightarrow{900^\circ K} AsCl_3 + HCl + HC{\equiv}CH + As + C_2H_6 + C$$

Under proces conditions, the acetylene is doubtless reduced to ethane and the $AsCl_3$ to elemental arsenic.

This method ensures reliable destruction of Lewisite. After one cycle of reaction, it is possible to obtain metallic arsenic of extra high purity. The rest of the reaction products are easy to utilize. However, because of the high reaction temperature and the presence of toxic agents mixed with hydrogen and hydrogen chloride in the reaction zone, this process is very unsafe for working staff and detrimental to the environment.

2.1.8. *Ammonia Reduction.*

Lewisite is reduced to elementary arsenic by reaction with ammonia up at temperatures of 1000-1100°C. Besides elementary arsenic, the products are non-toxic compounds, such as methane, nitrogen, and ammonium chloride.

$$ClCH = CHAsCl_2 + 6\ NH_3 \rightarrow As + 3\ NH_4Cl + 1.5\ N_2 + 2\ CH_4$$

$$AsCl_3 + 4\ NH_3 \longrightarrow As + 3\ NH_4Cl + 0.5\ N_2$$

This method has value for production of arsenic of high quality. However, the high temperature of reaction and use of a large quantity of ammonia pose certain dangers.

In summary, a great number of Lewisite destruction technologies have been considered competitively.

In course of the multistage technologies competition, much attention was paid to those directions that would permit utilization of arsenic for economic purposes.

In order to focus our efforts at one of those perspective directions in future, the Commander of Troops of Radiological, Biological, and Chemical Defence of Russian Federation Ministry of Defence requested an expert evaluation of alternative technologies with the aim of choosing the one most optimum with respect to ecological and technical safety.

Taking into consideration all the major factors mentioned, the experts recommended the technology of alkaline hydrolysis with subsequent electrolysis of the reaction products as the preferred technology for Lewisite destruction.

2.2 'MUSTARD' STOCKPILES DESTRUCTION TECHNOLOGIES

The favored technology for destruction of Russian stockpiled 'Mustard' was chosen after detailed research. The technology involves reaction of 'Mustard' with a mixture of monoethanolamine and ethylene glycol, with subsequent combustion of reaction masses (two stage technology).

Two Stage Technology of 'Mustard' Destruction. The reagent for the first stage comprises monoethanolamine and ethylene glycol in proportions 9 : 1. One kilogram of 'Mustard' requires one kilogram of the reagent. Reaction temperature is 100- 110°C; reaction time is about one hour.

The scheme of reaction may be presented as follows:

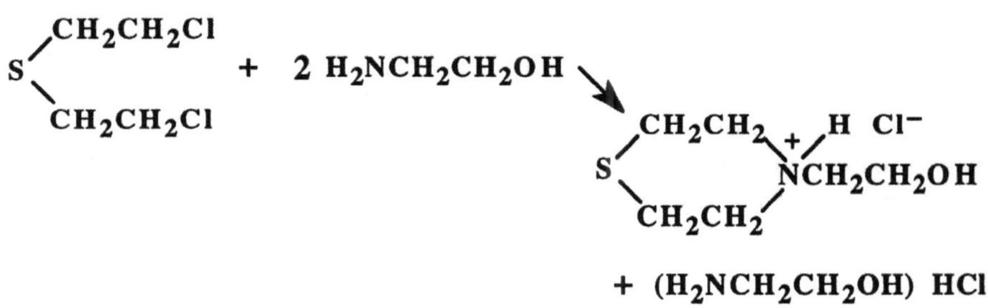

$$3\ S(CH_2CH_2Cl)_2 + 10\ HOC_2H_4OH \searrow$$

$$6\ ClCH_2CH_2OH + 6\ H_2O + S(C_2H_4OC_2H_4OC_2H_4SC_2H_4OC_2H_4OH)_2$$

The complicated mixture of reaction products should be burnt. Combustion of such mixtures may have dangerous ecological consequences, and produce dioxins. Therefore further research was undertaken to exclude combustion from the technology and to substitute for it another way of processing the reaction masses.

At the end of the research work, in place of combustion there was proposed to bitumenize reaction masses and subsequently to bury the bitumen as monolithic blocks in special burial grounds. Tough road bitumen with softening temperature 65-70°C was used in bitumenization.

Reaction masses were processed in bitumen with calcium hydroxide at temperature 170°C. The time of mixing was 45 minutes. The bitumen masses produced had softening temperature 81.5° C, which permitted forming them into monolithic blocks for subsequent burial.

Thus, the two stage technology for 'Mustard' destruction has much promise. Furthermore, research on 'Mustard' utilization showed that "pure" 'Mustard' can be easily converted into products having economic value. For example: half-sulphide caoutchoucs (thiokols). Thiokols are used in aviation, automobile, and building industries in gaskets and gasoline-resistant synthetic rubbers.

'Mustard' oxidation in the presence of catalyst can produce supersorbents of metals. They are used for metallurgy purposes and for extraction of heavy metals from sewage waters.

2.3 'MUSTARD'-LEWISITE MIXTURES DESTRUCTION.

During long term storage, 'Mustard'-Lewisite mixtures undergo profound change. Tough polymeric products, which may amount to 33% of each mixture, are formed. There are substantial differenes between mixtures of different provenance. Therefore, it is very difficult to choose a method universally suitable for destruction of these chemicals. Russian research on destruction of 'Mustard'-Lewisite mixtures has revealed two technologies of special interest.

Sulphurization in the Presence of N-Methylpyrrolidone. Interaction of 'Mustard'-Lewisite mixtures with sulphur at temperatures of 140-150°C is the essence of the process. A polymeric alloy, which softens at temperature 90-100°C, is produced. The resulting mass should be buried in special burial grounds.

Polymerization with Ethylene Glycol Monomethacrylic Ester and Monoethanolamine. This interaction (equations for reactions of Lewisite with this ester and of 'Mustard' with monoethanolamine are presented above) is carried out in presence of a hydrogen chloride acceptor. Azo-bis-isobutyronitrile is used as polymerization initiator. Duration of detoxification is 2-2.5 hours.

According to analyses that determine the residual content of 'Mustard' in reaction mixtures, the second technology is the more effective.

In conclusion, Russia has developed and refined methods for destruction of blister chemicals that are effective and often afford useful products. One can say with confidence, taking into account our experience and the experience of other states, that acceptable solutions of technological problems will be achieved.

3. Provisions For Ecologically Safe Destruction of Stockpiled Lewisite and 'Mustard'

In current research and development on destruction of CW agents and munitions in CW stockpiles, safety and protection of the environment are requirements that must be satisfied.

Practice shows that elaboration of systems for ecological preservation must, after care for human safety, take priority over other factors in developing an industrial technology. Commonly, such systems must include evaluation of environmental conditions, forecast of environmental impacts, documentation of any ecological effects, and ecological control of processes. In practice, in the course of the direct realization of CW destruction technologies, the principal considerations are evaluation of environmental conditions and monitoring of CW destruction processes to ensure that they are ecologically benign. As a rule, the forecast of environmental impacts and documentation of ecological effects are performed in the course of productions projection.

Consequently, side by side with other tasks not less important, the main feature of ecologically acceptable CW destruction is reduction of the ecological burden on the environment. This presupposes minimization of the number of toxic chemicals used in technological cycles of CW destruction and careful monitoring for identification of them.

For these purposes, systems of chemical analysis are used for ecological monitoring of the environment and ecological control over CW destruction. The appropriate analytical methods are defined by both the physico-chemical properties of chemicals and the character of the environmental factor of interest.

Accordingly, in our opinion, among many other factors contributing to ecologically safe destruction of blister chemicals, it is expedient to pay especial attention to two of major significance that are the technical foundation for ecologically benign destruction of 'Mustard' and Lewisite. The first is development of reliable, highly effective methods for purification or neutralization of toxic materials used in 'Mustard' and Lewisite detoxification. These methods should then be used in a technological destruction plan. The second is development of chemical analysis methods and equipment for determination of microconcentrations of blister chemicals and toxic products from their degradation in locations where destruction is performed, and at selected places in the environment. ("Chemical analysis methods" implies sampling, sample preparation, and analysis.) Such methods are features of Russia's technological scheme of CW destruction, with its system for ecological monitoring of the environment.

To save time, this Chapter includes only a few major technological aspects of the issues discussed, ones that we consider to be of major significance. In choosing purification and neutralization methods for CW destruction, our state is now paying more and more attention to the biological reprocessing of toxic chemicals, often enabling CW destruction to be performed in ecologically safe fashion. In recent years, much research work has involved searching for micro-organisms that can destroy blister chemicals, or genetic engineering to improve strains already at hand. In the course of this research, there were received about 700 such strains.

Experiments in the sphere of genetics show that tolerance to arsenicals is stipulated by the presence of plasmid with 2.4-87 MgD (Mega Daltons) molecular weight in cells of micro-organisms. We have also found strains of micro-organisms that cannot grow in media without arsenic, as well as strains that use thiodyglycol (a 'Mustard' destruction product) as their only source of carbon. In course of incubation of the latter with reaction masses from 'Mustard' alkaline hydrolysis, complete mineralization of thiodiglycol took place.

In the course of development of microbiological methods for purification of effluent waters from Lewisite destruction, straightforward schemes for purification of waters containing from 100 mg/m^3 to 10 g/m^3 of arsenic were worked out.

Use of highly active micro-flora especially adapted to high concentrations of arsenic in such biotechnological schemes is a new step.

Not only sewage waters, but also soils polluted by Lewisite, 'Mustard', and products of their destruction were exposed to microbiological remediation. It was determined that the velocity of the mentioned chemicals' decomposition was three times higher in samples with active micro-flora than in samples with natural micro-flora.

The main technical requirement during realization of the second factor for ecological protection, namely, development of chemical analysis methods and equipment, is high sensitivity. The necessary level of sensitivity is determined by the maximum permissible concentrations of toxic chemicals.

Thus, both methods and equipment for chemical analytic control, elaborated for the purpose of ecologically benign 'Mustard' and Lewisite destruction, must meet a high standard, namely, sensitivity, which is determined by the listed maximum permissible concentrations of toxic chemicals. At the same time, in the framework of each specific technology for detoxication of a specific chemical, it is necessary to perform research to determine what harmful and toxic chemicals may be produced in the technological process and what effects they may have on the environment It is also necessary to determine what are the maximum permissible releases of those chemicals.

4. Conclusion

The Government of the Russian Federation, the Russian Ministry of Defence, and, particularly, the Commander of Troops of Radiological, Biological, and Chemical Defence Department, are responsible for CW destruction in our state, and for protection of inhabitants who may be affected by CW destruction problems.

Special attention in our activity in this sphere will be paid to Russia's CWC obligations. Understanding that possible affects of any activity related to CW destruction may have a harmful influence on people's health and on the ecosphere, we are compelled thoroughly to evaluate, to solve problems of waste discharge, and in other respects also to perform control over CW destruction technologies. These requirements are predominant in development of CW destruction technologies.

In spite of the current hard economic situation of our state, the Russian Federation remains committed to the principle of general and complete chemical disarmament and will strain every effort to save the human race from such weapons of mass destruction, now and forever.

5. References

1. Russian President Edict *"On the Preparation of Russian Federation to the International Obligations Dischargement in the Sphere of Chemical Disarmament"* **314**, 24 March, 1995.
2. Petrov, S.V. (1996) Chemical weapons destruction technologies are chosen. Pilot plants construction is begining, *J. Nuclear Control* **13**, 2-6.
3. Umjarov, I.A., Kuzneczov, B.A., Kholstov,V.1. (1993) Lewisite and 'Mustard' destruction (utilization) methods, *Russian Chemical Magazine* **3**, v. XXXVII, 25-29.
4. Shantroha, A.V., Gormay V.V. (1994) Chemical-Technological provisions of problem of blister agents destruction (utilization), *Russian Chemical Magazine* **2**, v. XXXVIII, 23-25.

RECOVERED OLD ARSENICAL AND 'MUSTARD' MUNITIONS IN POLAND

Technologies, Plans and Problems

ZBIGNIEW WERTEJUK, MIECZYSLAW KOCH,
WLODZIMIERZ MARCINIAK
Military Institute of Chemistry and Radiometry
00-910 Warsaw, Poland

1. Outline of chemical weapons on the territory of Poland

The history of chemical weapons (CW) may be divided into four periods: World War I, the interwar period, World War II and the period after World War II.

The first mass use of CW agent by the Germans against the Russians during World War I took place on Polish territory [1]. T-Stoff, a mixture of brominated aromatic hydrocarbons, mainly containing xylyl bromide, was used in shells for heavy howitzers (caliber 150 mm) on the Rawka River front by Bolimów on 30 January 1915. Due to unfavourable weather (temperature below –20 °C), the concentration of T-Stoff vapour was so low that the Russians did not even notice its presence. The next use of CW agent took place almost in the same place during the night 30/31 May 1915. The Germans collected 12,000 gas cylinders, containing 264 tons of chlorine, along the 12 km front. In this case the Russian casualties were very high: 8934, of whom 1101 died. A week later, during the night 6/7 June 1915, the Germans carried out their third chemical attack in the same place, using also chlorine gas cylinders. When the cylinders were discharged, the wind turned round and blew the chlorine cloud towards the German trenches, causing them casualties too.

During the interwar period independent Poland (1918 - 1939) signed the Geneva Protocols for the Prohibition of the Use in War of Asphyxiating Poisonous or Other Gases, and Bacteriological Methods of Warfare on 12 June 1925 and ratified them on 4 February 1929. Research on protection against CW and investigation of the properties of CW agents (e.g., fluoroorganic compounds, sulfur 'mustard' and Lewisite) were carried out [2]. At that time, also, a small amount of sulfur 'mustard' was manufactured in Pionki (armament plant) mainly for training purposes. It was used for filling

J.F. Bunnett and M. Mikołajczyk (eds.), Arsenic and Old Mustard: Chemical Problems in the Destruction of Old Arsenical and 'Mustard' Munitions, 91–104.

artillery shells and land mines as a mixture containing 5 - 15% of 'mustard'. The total capacity of each mine was approximately 12 dm^3, whereas its work capacity was approximately 9 dm^3.

During World War II, CW agents were used also on the territory of Poland. On 3 September 1939, German aircraft dropped a few bombs filled with sulfur 'mustard' on one of the suburbs of Warsaw. According to German military authorities it had been used by mistake. On the other hand, there is information from German sources that Polish Armed Forces laid "mustard gas" and fired mixed chemical barrages against adversary troops near Jaslo. In Polish sources there is no evidence to confirm that report. During World War II, CW agents were manufactured by Germany mainly in a factory at Dyhernfurth on the Oder, at present named Brzeg Dolny, constructed in 1940. Prisoners were working in the factory where CW agents (especially tabun) were manufactured. Prisoners were also placed in gas chambers for testing the influence of tabun upon human beings. Monthly production of tabun amounted to 1000 tons and probably totaled 12,000 tons in 1940 - 1945. This agent was used for filling 105 and 150 mm artillery shells, 250 kg aviation bombs and storage tanks. In this same plant, the Germans also produced a large amount of hydrogen cyanide and cyanogen chloride (approximately 20 tons/month). Near the end of World War II, 9.3 tons of German Adamsite was abandoned on Polish territory. This agent still awaits destruction.

After World War II a few tons of sulfur 'mustard', several kilograms of Lewisite and an unknown amount of chloroacetophenone were manufactured at Brzeg Dolny. Among them only chloroacetophenone was used: to fill hand grenades and smoke candles for training purposes. Nowadays no production of CW agents takes place at Brzeg Dolny. The installations for their production have been dismantled and no longer exist. At present there is a chemical factory named Rokita, for organic chemical production, at Brzeg Dolny.

At research laboratories of the Military Institute of Chemistry and Radiometry and at the Military University of Technologies, a few kilograms of organophosphorus warfare agents (tabun, sarin, soman, VX) and some amount of CS were synthesised. None of these warfare agents was produced for use during a military conflict. They were used for investigations, training of soldiers and checking the effectiveness of equipment for protection against CW agents. Nearly all of those warfare agents was used for training purposes or destroyed. At present only some amount of riot control agents (chloroacetophenone and CS) is still stored.

On Polish territory, various chemical munitions and CW agents are still found as relics of World War I and II. These are a few examples:

◆In 1956, chemical shells and mines containing "mustard gas" were found on the Baltic Sea beach at Darlówek. The unexpected 'mustard' contaminated 120 children from summer camps.

◆In 1959, a tank with 60 m^3 of organoarsenic chloride compounds was discovered on a dune at Jaroslawiec.

◆In 1970, during earth work, 32 mines including 16 filled with "mustard gas" were unearthed on a schoolground at Przemysl.

◆At Ilawa, some chemicals were found on the site of former German barracks during excavations. They were spray canisters with remnants of an unidentified liquid and stoneware containers with xylyl iodide.

◆In 1991, at Wizna, two barrels of sulfur 'mustard' were found in an area of former fortifications.

A separate problem is that of chemical munitions sunk in the Baltic. These munitions were produced during World Wars I and II, and after World War II; they were sunk by Weimar Germany, the former GDR and the former Soviet Union. In the Polish economic zone, the sunken chemical munitions are located in 5 regions. It is estimated that the total area of those 5 regions is a few thousand km^2.

Those sunken munitions are a serious hazard. "Mustard gas" contamination of beaches near Kolobrzeg, Mielno, Darlówek, Jastarnia and Hel have been recorded, as well as injury to fish cutters from contaminated nets. Moreover, there have been five incidents in which artillery shells and bombs with "mustard gas" were cast upon the beach by the Baltic Sea, with consequent contamination of 126 people. At sea, on 13 occasions fishermen have been harmed by exposure to "mustard gas" on their gear.

2. Some problems in dealing with CW agents, and Polish contributions to their solution

Two contributions merit special attention:

- ◆ a two-stage method for destruction of a few tons of sulfur 'mustard'
- ◆ methods for destruction of 9.3 tons of Adamsite abandoned in Poland

2.1. A TWO-STAGE METHOD FOR SULFUR 'MUSTARD' DESTRUCTION

In selection of a technology for destruction of a CW agent, one must take into account the quantity and type of toxic agent, the efficiency and safety of the process, protection of the environment, and costs of training personnel.

In the case of Poland, which has no chemical weapons of its own production but has small quantities from other countries, relics of World War I and II, the economic base to construct a special installation for thermal destruction of CW agents does not exist. Desired therefore are methods for destruction of these modest quantities of CW agents that are specially crafted for each and employ relatively simple equipment, preferably in installations that already exist. Obviously, it is imperative that a chosen technology be safe for people and kind to the environment.

Taking into account all these conditions, a two-stage method for destruction of "mustard gas" has been chosen [3,4]. The initial stage is chemical detoxification of the agent by reaction with an alcoholic solution of sodium hydroxide. Methyl alcohol was selected as solvent because of the following factors:

the accessibility and low cost of technical methanol

the relatively high solubility of polymeric 'mustard' contaminants in methanol,

the prospect of using the decontaminating agent as fuel in the second (combustion) stage of the process.

The chemical destruction of pure sulfur "mustard gas" by reaction with a methyl alcoholic solution of sodium hydroxide is represented in part by the following equations:

$$S(CH_2CH_2Cl)_2 + 2\ NaOH \longrightarrow S(CH_2CH_2OH)_2 + 2\ NaCl$$

$$S(CH_2CH_2Cl)_2 + 2\ NaOH \longrightarrow HOCH_2CH_2SCH{=}CH_2 + 2\ NaCl + H_2O$$

Side reactions form these by-products:

2-methoxyethyl vinyl sulfide $CH_2{=}CH{-}S{-}CH_2{-}CH_2{-}O{-}CH_3$

2'-hydroxyethyl 2-methoxyethyl sulfide $HOCH_2{-}CH_2{-}S{-}CH_2{-}CH_2{-}O{-}CH_3$

bis(2-methoxyethyl) sulfide $CH_3{-}O{-}CH_2{-}CH_2{-}S{-}CH_2{-}CH_2{-}O{-}CH_3$

The first stage, of "mustard gas" detoxification, took place in a cubical steel tank of 2.5 m^3 capacity (150 kg HD; 450 dm^3 solution of sodium hydroxide in methanol). The tank was closed at the top by a packed steel cover. In the cover there was a funnel that facilitated charging the tank with HD and the decontaminating agent used for flushing of the HD barrel. On completion of charging, the funnel was closed by an additional seal. Fitted to the reactor tank were a heat exchanger (not used, however, because sufficient heat was generated by the chemical reaction) and a centrifugal pump of 15 m^3/h capacity, used for mixing of reactor contents during the process as well as for pumping of reactor contents into an intermediate tank, from which reaction products were sent to combustion. For safety, the reactor was surrounded by a steel barrier wall to trap reactor contents should there be an accident.

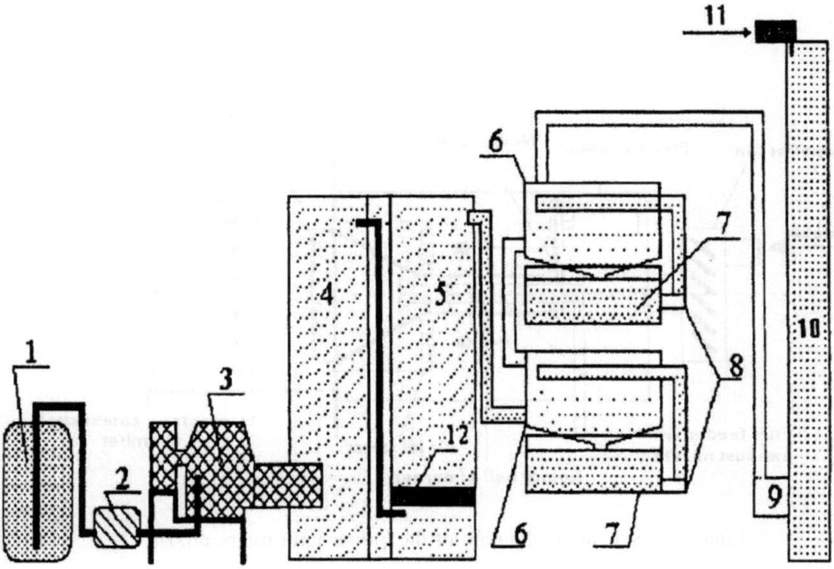

1 - intermadiate tank; 2 - force pump; 3 - burner with rotary-nozzle; 4 - elementary combustion chamber; 5 - recombustion chamber; 6 - scrubber; 7 - absorbate tank; 8 - absorbate pump; 9 - chimney fan; 10 - chimney; 11 - flue gas analyser; 12 - grate with solid fuel (coke).

Figure 1. Schematic drawing of installation for combustion of "mustard gas" detoxification products.

Time of the reaction was approximately 10 hours. After completion of reaction and analytical verification that no "mustard gas" remained, the mixture was pumped into intermediate tanks made of polypropylene.

The second stage of the process for destruction of "mustard gas" involved combustion of the detoxified products in a special installation (Fig. 1). The combustion process was conducted periodically after accumulation of an adequate quantity of detoxification products in intermediate tanks. In accordance with a principle of a work of installation, the solution containing detoxification products was burnt in two stages in two chambers arranged one after the other. In the first chamber, the combustion was realised with use of a specially designed burner with rotary-nozzle (Fig. 2), which allowed burning of fuel having solid impurities. The burner was characterised by small burning dynamics (1.5 dm^3/min) giving a guarantee of getting maximal applicable temperature.

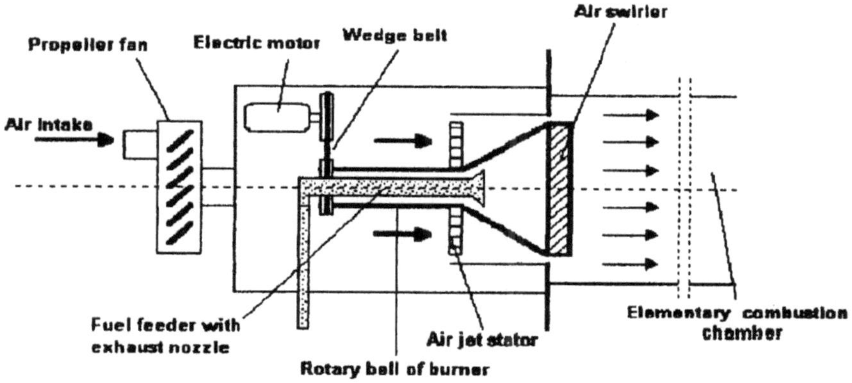

Figure 2. Schematic diagram of the burner with rotary nozzle.

2.2. ADAMSITE DESTRUCTION

From a chemical point of view Adamsite is diphenylaminochloroarsine (phenarsazine chloride) with standard NATO agreement code DM.

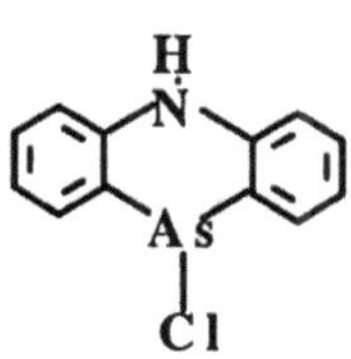

Adamsite

Adamsite belongs to the riot control agent family, being of vomiting agent type. The properties of Adamsite are as follows:

density 1.65 g /cm^3 at temperature 20 °C (293 K)

melting point 195 °C (468 K)

boiling point (with decomposition) 410°C (683 K)

As an aerosol, Adamsite severely irritates the eyes and respiratory tract, causing tears, sneezing and coughing

median lethal dosage (Lct_{50})) 15,000 mg-min.m^{-3}

median incapacitating dosage (Ict_{50})

(one minute) 22 mg-min m^{-3}

(one hour) 8 mg-min

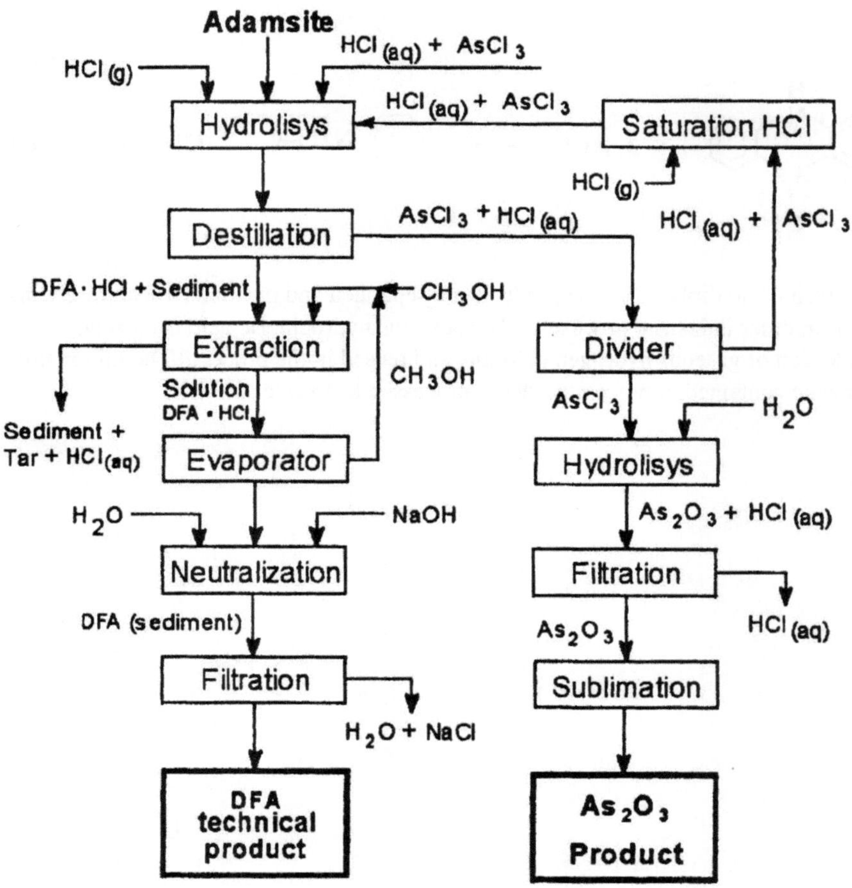

Figure 3. Adamsite destruction by using hydrochloric acid.

Adamsite was discovered in 1918 by an American chemist Adams (after whom it is named). It was produced on industrial scale in several countries (among others USA, Italy, Great Britain, Germany) until the 1930's. Its production was given up after development of an industrial method for chloroacetone production. The latter has irritating properties similar to those of Adamsite but lacks the poisonous properties characteristic of arsenic compounds.

Destroying the Adamsite is an urgent necessity. Its destruction is however expensive and fundamentally difficult. The main problem is that even after burning or chemical conversion of Adamsite, the unavoidable arsenic-containing products are still toxic. On the basis of theoretical studies and laboratory works performed at the Military Institute of Chemistry and Radiometry and at the Military University of Technology, the methods described below have been proposed [5,6].

2.2.1 *Adamsite undergoes hydrolysis in concentrated aqueous hydrochloric acid, according to the equation* [6]:

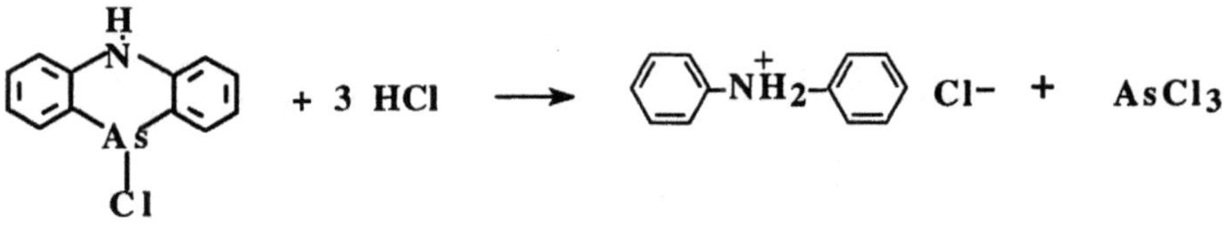

The arsenic and diphenylamine products are separated and purified to become chemicals useful in other industrial proceses. The aqueous hydrochloric acid is cleaned, saturated by addition of gaseous hydrogen chloride, and reused in the process. The minor amounts of arsenic contained in process wastes is processed to arsenic sulfide.

2.2.2. *Adamsite fusion with sulfur* (Fig. 4)

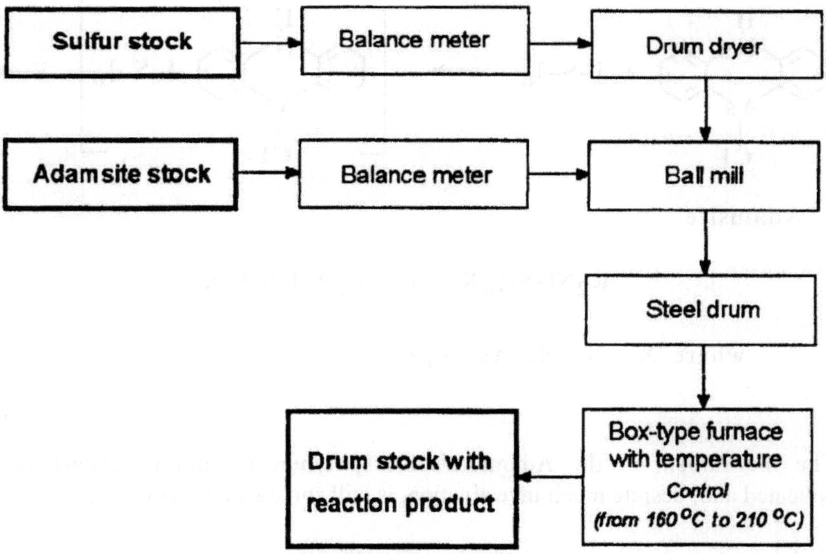

Figure 4. Adamsite destruction by using sulfur.

The fusion of Adamsite with sulfur is effected in hermetic vessels [7]. The weight ratio, Adamsite to sulfur, is from 1:4 to 1:5. In the early phase of the fusion process, the temperature is maintained at 160°C for 2 hours; it is then raised to 200°C for one hour. The Adamsite/sulfur fusion process, which furnishes a solid reaction product, probably occurs according to the scheme:

(a) formation of sulfur radicals

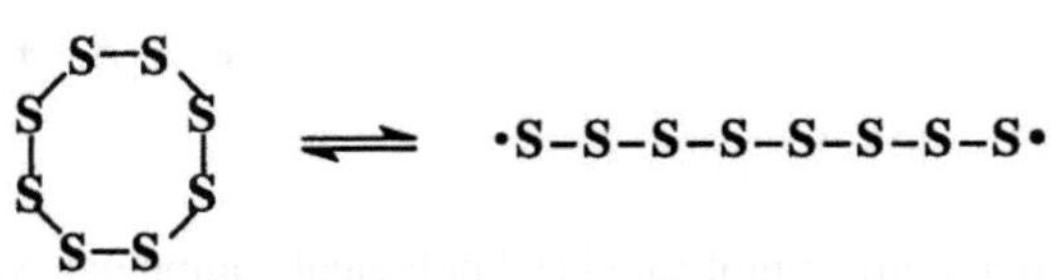

sulfur (S_8)

(b) sulfur chains propagation

$$\bullet S{-}S{-}S{-}S{-}S{-}S{-}S{-}S\bullet \;+\; (n-1)\,S_8 \longrightarrow ([-S-]_8)_n$$

(c) reaction with Adamsite

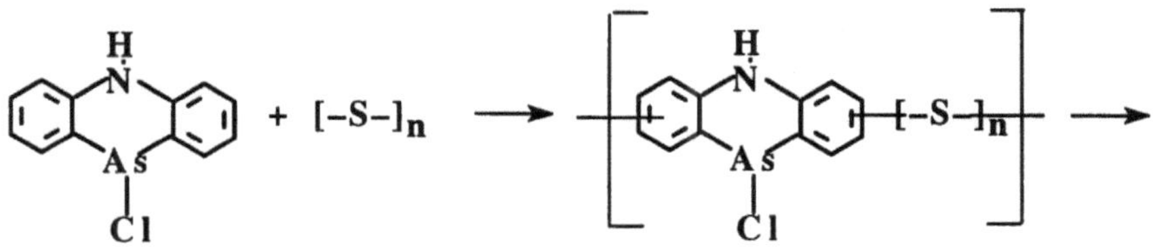

Adamsite

$$\longrightarrow \mathbf{R_2N[\text{-}S\text{-}]_nX} \quad \text{or} \quad \mathbf{R_2As[\text{-}S\text{-}]_nX}$$

where $\mathbf{X} = \text{-S, -N, -As, -}C_6H_5$

The mechanism of the Adamsite/sulfur polymer formation process is very complicated and, despite much investigation, is still somewhat speculative.

2.2.3. *Adamsite reduction by phosphorous acid* (Fig. 5)
Adamsite reacts readily with phosphorous acid in concentrated acetic acid solution or directly in fused phosphorous acid [6]. This method is easily realised in an organic chemical production plant. The reaction is irreversible and leads directly to useful products: metallic arsenic of high purity, and diphenylamine hydrochloride.

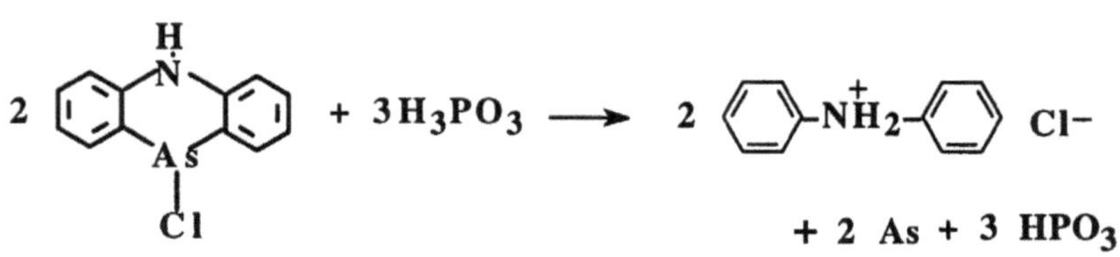

The process does not require complicated technological equipment. Separation of arsenic is by filtration and diphenylamine is obtained by crystalisation.

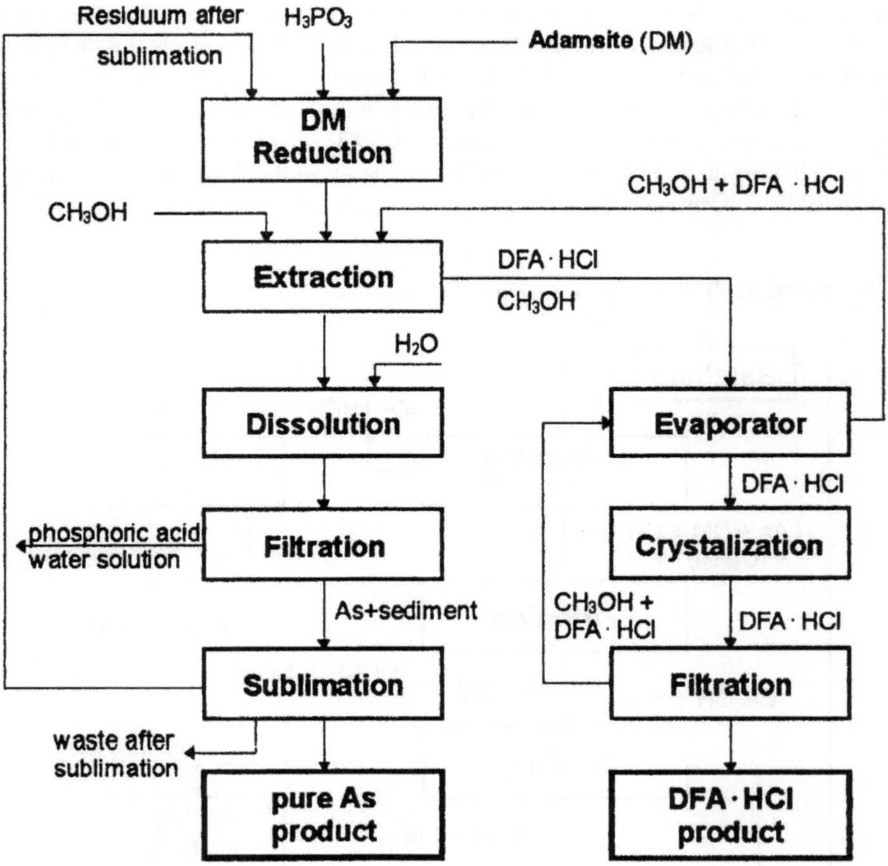

Figure 5. Adamsite destruction by using phosphorous acid.

2.2.4. *Adamsite reduction by formic acid* (Fig. 6)

Formic acid destroys Adamsite according to the scheme [5,8,9]:

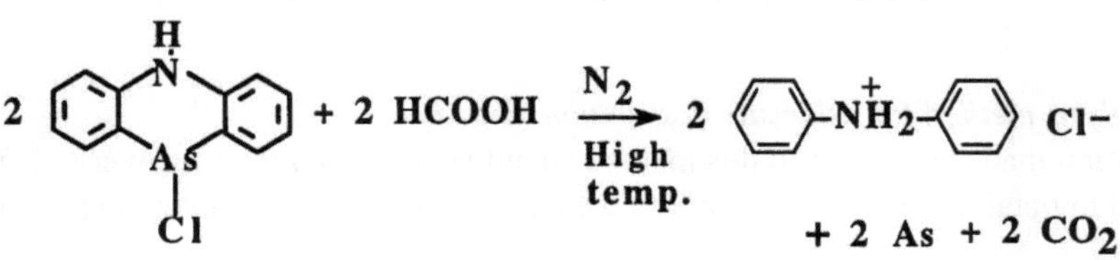

In laboratory tests it was found that this method is unprofitable for large-scale destruction of Adamsite. Information in the literature [10] suggests that the process yield and efficiency may be improved when reaction conditions are more severe. Process should proceed in the special shunting kiln in steam - gas phase in the presence water steam and of hydrogen at temperature of 900 - 1000°C. Arsenic trichloride, arsenic, hydrogen chloride and soot, which spring up in the Adamsite destruction, are separated by condensation, filtration and absorption processes.

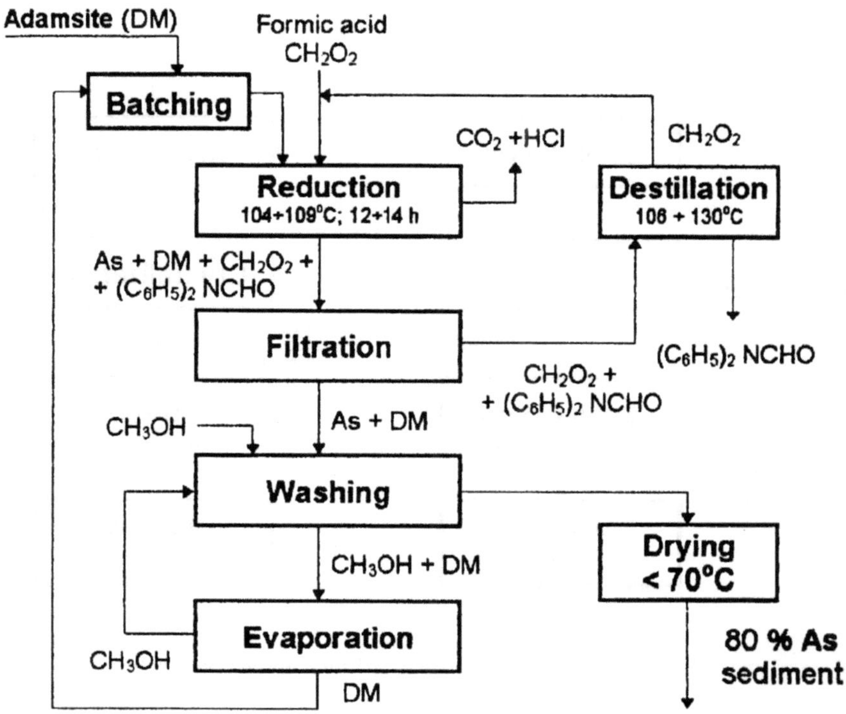

Figure 6. Adamsite destruction by using formic acid.

2.2.5. *Pyrolysis method for Adamsite destruction* (Fig. 7)

An installation made according to this method shoud lead to maximum recovery of free arsenic and to obtaining wastes harmless to the environment from the remaining arsenic compounds.

The main installation units would be: sublimation and pyrolysis chambers and a cooler in the form of a slab heat converter. Adamsite will be introduced into the sublimation chamber and transformed into vapour at the temperature of 400°C. The Adamsite vapour transported by inert gas (e.g., nitrogen), will pass into the pyrolysis chamber and will be decomposed to free arsenic, hydrogen chloride, nitrogen, low molecular weight hydrocarbons and soot. Free arsenic will be precipitated in the cooler from which it will be removed after partial disassembling of this unit. Units to bind hydrogen chloride, to catch the rest of the arsenic, and to burn hydrocarbons catalytically will be further parts of the installation. It is established that the sublimation and pyrolysis chambers and the cooler will work periodically, with some "down time".

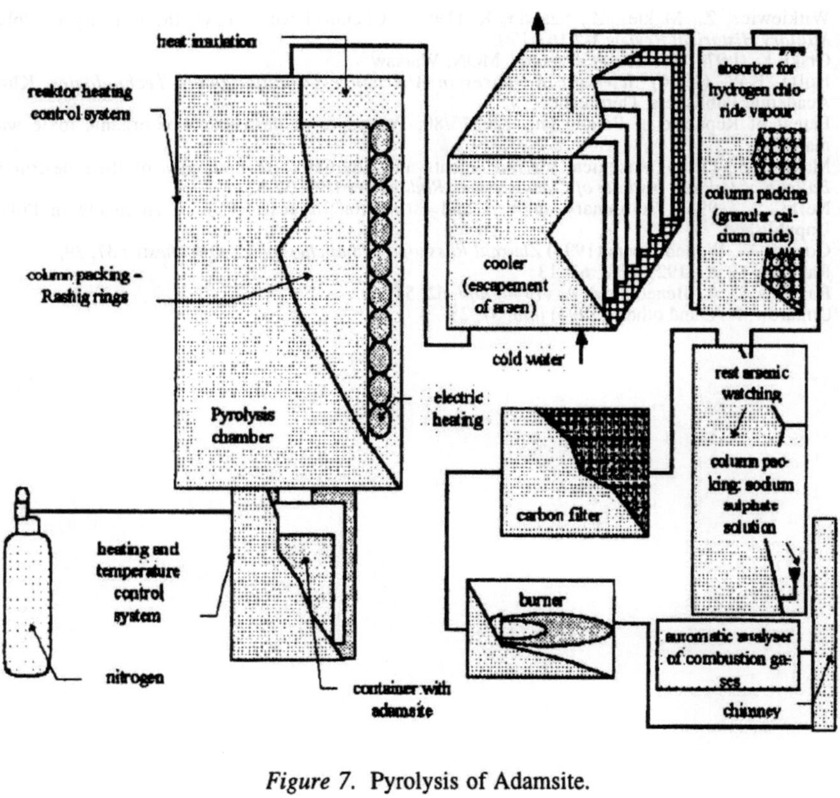

Figure 7. Pyrolysis of Adamsite.

3. Conclusions

◆The two-stage method is an efficient way for sulfur 'mustard' destruction, harmless to people and the environment, especially useful for destruction of small amounts of the CW agent.

◆Among the methods for Adamsite destruction presented, the most convenient seem to be Adamsite pyrolysis and its fusion with sulfur.

4. References

1. Witkiewicz, Z., Makles, Z., Szarski, K. (1995) Chemical weapons on the territory of Poland, *Military Historical Review* **1-2**,163-170.
2. Orski, K. (I 974) *The Secret Arsenals*, MON, Warsaw.
3. Holm, F.W. (1996) *Scientific Advances in Alternative Demilitarization Technologies*, Kluwer Academic Publishers, Dordrecht.
4. Patent of Republic of Poland No. 126275/80, Procedure of destruction of organic toxic waste materials.
5. Makles, Z., (1995) Arsenical warfare agents and alternative technologies of their destruction, *Bulletin of Military Institute of Chemistry and Radiometry* 1/95, **37**-62.
6. Neffe, S. (1995) Preliminary analysis and estimation of destruction of Adamsite in Poland, Unpublished.
7. Gormai, W. W. and others (1994) *Zhurnal Russkogo Chimitcheskogo. Obshtchestva* **37**, 39.
8. Razujew, G. A. (1929) *ibid.* **61**, 13.
9. Razujew, G. A., Benediktow, A. (1930) *ibid.* **62**, 547.
10. Umiarow, I. A. and others (1993) *ibid.* **37**, 25.

COMPOSITION AND REMEDIATION OF TARRY 'MUSTARD'

P.R. NORMAN
Chemical and Biological Defence Sector
PLSD, Porton Down, Salisbury SP4 0JQ, U.K.

1. Introduction

Before considering the likely composition of "tarry mustard", it is pertinent to first dwell on the chemistry and reactivity of the material from which it is derived, i.e. bis(2-chloroethyl) sulfide (Figure 1) or "mustard gas".

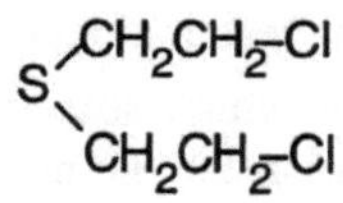

Figure 1

It is likely that the material referred to as "tarry mustard" will be of an almost indefinable composition; as varied as the wide number of formulations of 'mustard' that have been used in the past to fill munitions and the complex chemistry displayed by "mustard gas" in its pure form.

This paper will highlight these complex aspects of 'mustard' formulation and chemistry and will describe initial attempts to identify the chemical species present in one sample of "tarry mustard". The rich chemistry of "mustard gas" is too extensive to describe fully in this short article, and the examples cited within are merely examples of the different types of reactivity displayed by "mustard gas".

The presentation will conclude with a brief description of a process being developed in the UK that offers the potential of rendering the contents of old 'mustard' munitions safe, irrespective of the nature of the organic species that might be present in the munition.

2. 'Mustard' Formulations and Reactions

2.1 PRODUCTION OF 'MUSTARD'

"Mustard gas" was first synthesised in the middle of the nineteenth century. Niemann[1] and Guthrie[2] were amongst workers who published details of its preparation in 1860. In general, three synthetic routes for 'mustard' production were developed. The first of

J.F. Bunnett and M. Mikołajczyk (eds.), Arsenic and Old Mustard: Chemical Problems in the Destruction of Old Arsenical and 'Mustard' Munitions, 105–114.

these was the reaction of sulphur monochloride with ethene and is shown below in Figure 2.

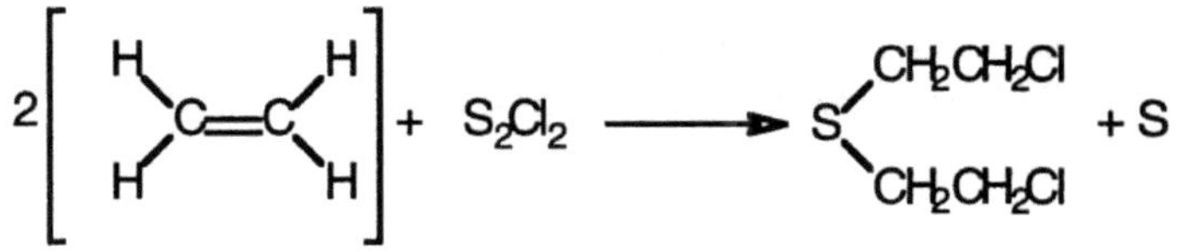

Figure 2

The major disadvantage with this process was the production of molecular sulphur which had to be removed at the end of the process. Carried out at low temperature, this reaction was the basis for the Levinstein Process which was extensively used for 'mustard' production during and after the First World War.

'Mustard' was also produced by the reaction of ethene with sulphur dichloride (SCl_2). The process took place in a solvent, which sometimes had to be removed from the reaction mixture at the end of the process, but did not produce sulphur as a by-product.

The third method was by direct chlorodehydroxylation of 2,2'-thiodiethanol (thiodiglycol, TDG). There are several ways of achieving this. In some cases, the material bis[2-(2-chloroethylthio)ethyl] ether is produced in a side reaction. Mixtures of this later material with 'mustard' have often been used to fill munitions in the past due to the favourable physical properties of the mixture.

The reaction between sulphur monochloride and ethene has been shown to produce a wide variety of by-products. The exact nature of these by-products is very dependent on the reaction parameters, but as an example, 'mustard' produced by this process commonly has melting points between 2 °C and 12 °C, an indication of the variability of the product. Sulphur produced during this reaction had to be removed, and this was often achieved by diluting the 'mustard' product with monochlorobenzene and separating out the sulphur after settling. In addition to quantities of chlorobenzene being found in 'mustard', additives such as iron, ethanol and pyridine have all been investigated as catalysts of the reaction. It is therefore not surprising that the "technical mustard" used to fill munitions was often far removed from the pure material on which so many laboratory based studies have been carried out. Materials such as hydrocarbons, polysulphides, ethene disulphide, bis(2-chloroethyl) disulphide and bis[2-(2-chloroethylthio)ethyl] ether have all been found in munition fills.

In addition to by-products resulting from the production of 'mustard', certain substances were deliberately added to "technical mustard" to bring about a wide variety of effects. The material bis[2-(2-chloroethylthio)ethyl] ether, which has already been mentioned, was often added to 'mustard' in a mixture which contained about 60% 'mustard'. This fill is often to be found in old munitions. Some countries used a mixture of Lewisite and 'mustard' as a weapon fill, and many countries added materials such as alloprene, methyl methacrylate, etc. to thicken it.

The net result of the many processes and variations that occurred with 'mustard' munition fills is that firstly, the munition was comparatively rarely filled with pure 'mustard', and secondly, the variations were so numerous that it is difficult to know what fills were originally in munitions, let alone what that material may have degraded to over the years.

2.2. REACTIONS OF 'MUSTARD'

'Mustard' is a very reactive material, and obviously, "tarry mustard" will come about as a result of this reactivity. The purpose of the few reactions described below is merely to illustrate the diverse chemistry of 'mustard'; there are many more examples of 'mustard' reactivity than those given within.

'Mustard' reacts with a wide variety of nucleophiles. Exemplified within this class of reactivity are reactions with hydroxide ion (hydrolysis) and reactions with sulphur and nitrogen nucleophiles. In addition, divalent sulphur is electron rich, and consequently 'mustard' is also attacked by electrophiles. Oxidation, chlorination and complexation typify this class of reactions. Recent studies have also explored the possibility of using biomimetic systems (catalase) to achieve facile oxidation of 'mustard' by peroxide moieties.

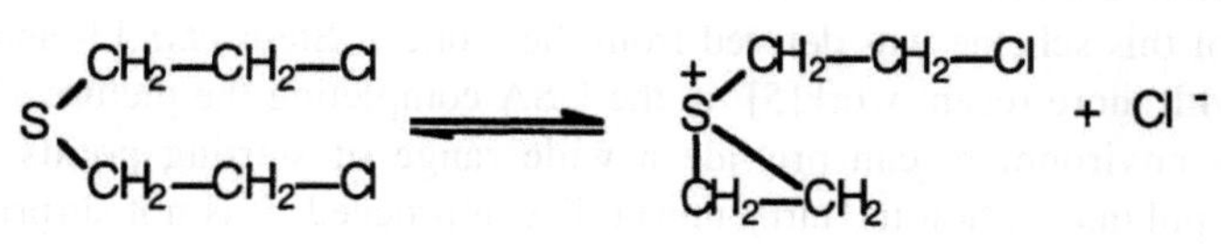

Figure 3

Much of the aqueous chemistry is dominated by the existence of an equilibrium between 'mustard' and a cyclic sulphonium ion (Figure 3). Hydrolysis is a result of hydroxide ion attack on a carbon atom of the three-membered sulphonium ring. The formation of the sulphonium ion is reversible and rate determining in the pH range 7-9. As might be expected, the rate of hydrolysis is inhibited by chloride ion and an increase in ionic strength enhances the rate of reaction as long as the ionic species do not participate in the reaction. If nucleophilic anions are present, they will compete with hydroxide ion and the rate of hydrolysis will decrease.

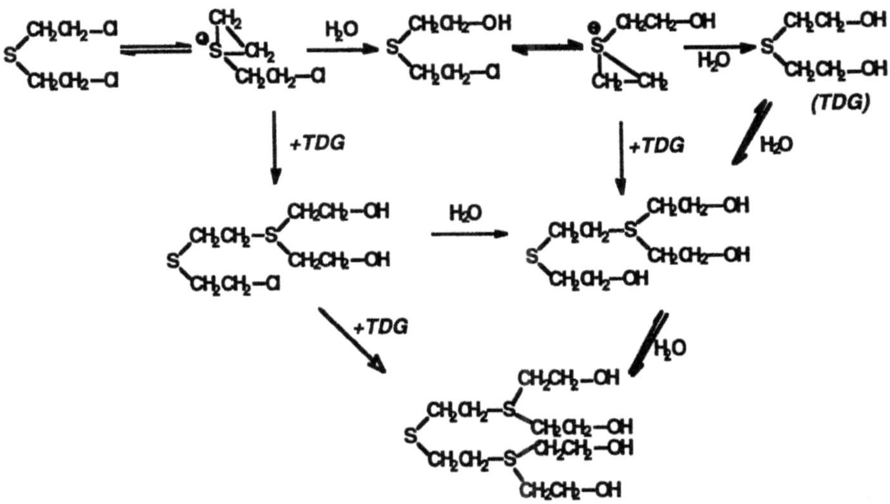

Figure 4

'Mustard' undergoes a complex series of hydrolysis reactions as is illustrated in Figure 4. Much of this scheme was derived from the work of Stein *et.al.*[3] and Bartlett and Swain[4] with more recent work[5] in the USA completing the picture. 'Mustard' in an aqueous environment can provide a wide range of starting points from which subsequent polymerisation to "tarry mustard" can proceed. It is not surprising that the exact nature of "tarry mustard" is not well defined.

Typical reactions of 'mustard' with nucleophiles are shown in Figure 5. Strong bases such as KOH will cause elimination of HCl from the material to yield divinyl sulphide. One equivalent of base will cause elimination in one 'mustard' sidearm resulting in a vinyl chlorethyl thioether which itself can undergo a whole series of hydrolytic equilibria. 'Mustard' will react with primary amines to produce cyclic products and will also react with higher amines and amino acids. 'Mustard' reacts with sulphide ion to produce a cyclic dithiane. This material is of interest since it is a useful analytical indicator of the former presence of 'mustard' in old environmental samples and munitions.

As already mentioned, nucleophilic anions react with 'mustard'. They can form either cyclic or disubstituted species. A wide range of activities have been observed which have been quantified in terms of a competition factor which provides a measure of the reactivity towards the 'mustard' sulphonium ion of the anion compared to hydroxide ion in water. Keeping in mind that concentration effects do need to be considered, Table 1 shows a small number of examples of competition factors which have been measured.

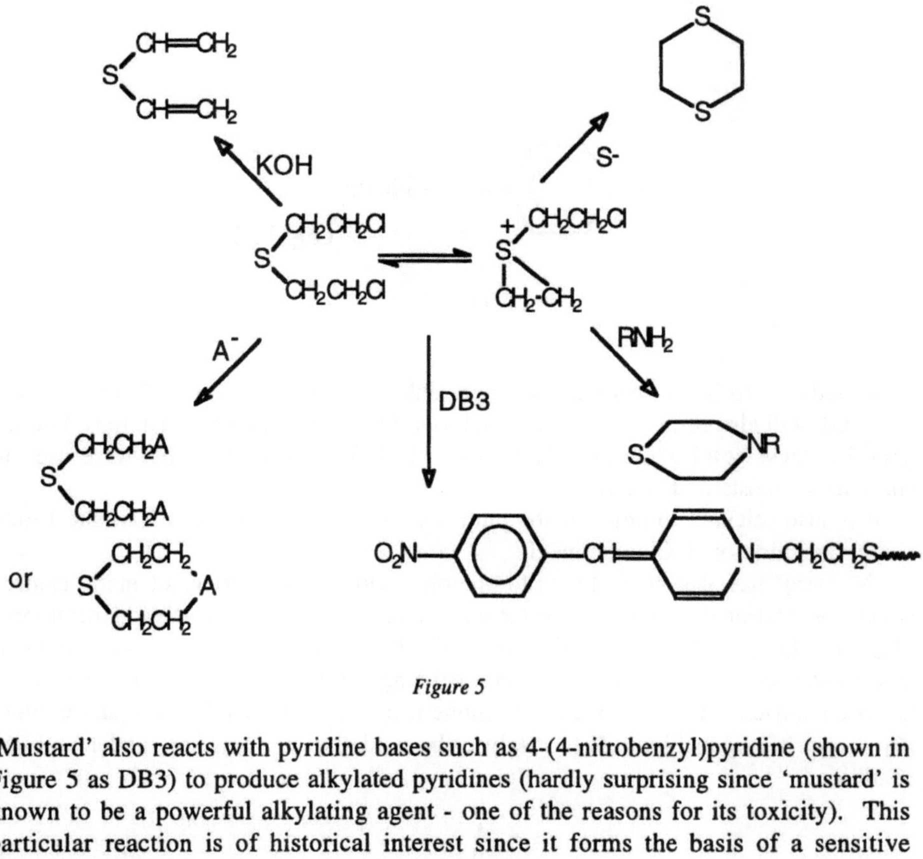

Figure 5

'Mustard' also reacts with pyridine bases such as 4-(4-nitrobenzyl)pyridine (shown in Figure 5 as DB3) to produce alkylated pyridines (hardly surprising since 'mustard' is known to be a powerful alkylating agent - one of the reasons for its toxicity). This particular reaction is of historical interest since it forms the basis of a sensitive colourimetric method for the analysis of "mustard gas".

Table 1. Examples of Competition Factors for 'Mustard' Reactions

Anion	Competition Factor
Dithiophosphate	130,000
Monothiophosphate	38,000
Thiosulphate	27,000
Methylamine	390
Chloride	21
Glycine	1.6

'Mustard' will also react with a wide variety of electrophiles. Its reactions with chlorine containing reagents such as hypochlorite ion are many and varied. Depending upon conditions,'mustard' can be oxidised to HCl, CO_2, SO_2, RCl and S (e.g. by solid $Ca(OCl)_2$) or to sulphoxide (sulphur IV) and sulphone (sulphur VI) species by hypochlorite ion in aqueous media.

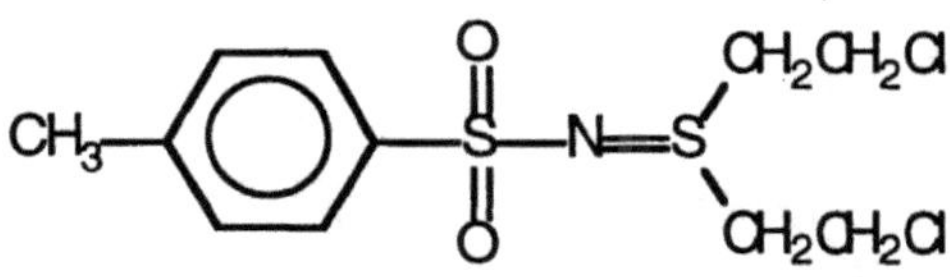

Figure 6

In addition to being oxidised by active chlorine reagents such as Chloramine-T, 'mustard' will also form sulphimine complexes (Figure 6) with these materials. In days gone by, these solid white materials with well defined melting points were used to characterise 'mustard' derivatives.

It is also relevant to note for the purposes of this workshop that Lewisite forms similar arsimines with Chloramine-T.

'Mustard' has also been shown to complex with a wide variety of metal centres which are electron deficient, such as the incomplete 'd'-orbitals of transistion metal ions. Mg, Sn, Hg, Bi, Cu, Ni, Ti, Cr, Mo, Cd, Pt, Zn, Fe and others have all been investigated in the past as means of immobilising "mustard gas". More recently work has been carried out at PLSD using biomimetic systems such as the manganese dimer [6] shown below in Figure 7 to catalyse the oxidation of 'mustard' under ambient conditions.

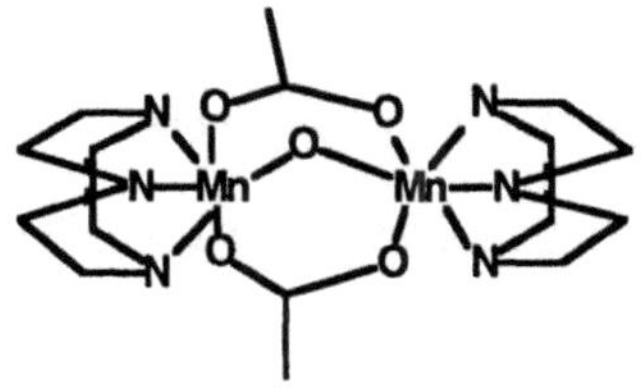

Figure 7

3. Composition of "Tarry Mustard"

The fact that 'mustard' can undergo such a wide variety of reactions as well as originating from a wide number of formulations would suggest that old samples are likely to contain a very large number of molecular species.

Figure 8 shows the analysis of mobile liquid found in a sample of 'mustard' suspected to be of World War II origin. The analysis shows seven significant materials in the mixture which were identified by gas chromatography with mass selective detection (MSD). Similar analyses from the vapour above a sample of old "mustard

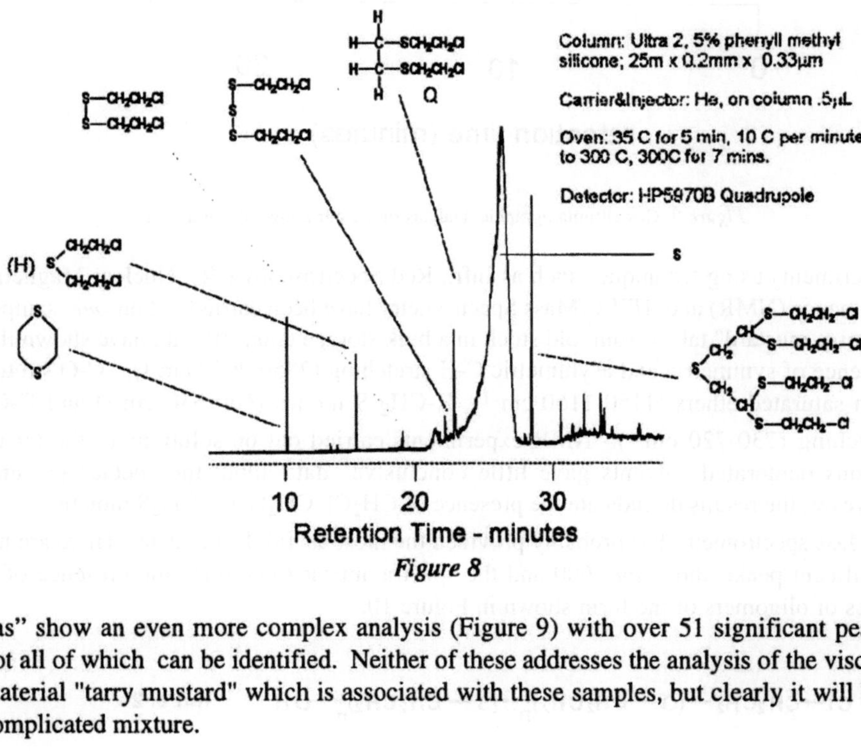

Figure 8

gas" show an even more complex analysis (Figure 9) with over 51 significant peaks, not all of which can be identified. Neither of these addresses the analysis of the viscous material "tarry mustard" which is associated with these samples, but clearly it will be a complicated mixture.

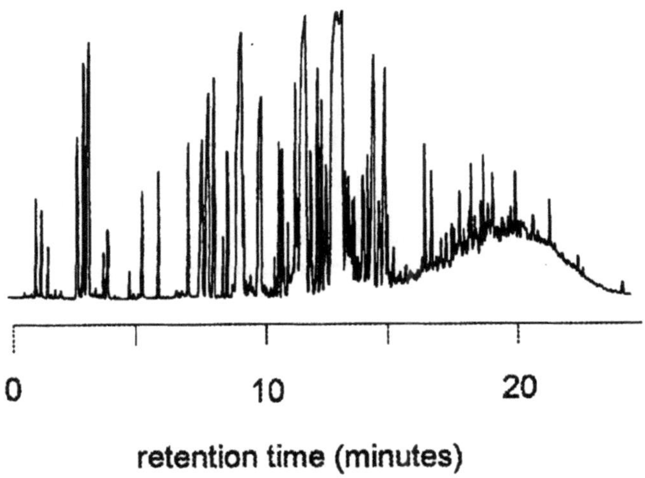

Figure 9 Gas chromatographic analysis of vapour above old 'mustard'

Experiments using techniques such as Infra-Red Spectroscopy (IR), Nuclear Magnetic Resonance (NMR) and HPLC/Mass Spectrometry have been carried out on *one* sample of "tarry mustard" taken from old stock in a bulk storage tank. IR data have shown the presence of symmetric and asymmetric C-H stretching (2850-2930 cm^{-1}), C-O stretch from saturated ethers (1150-1160 cm^{-1}), R-CH_2-S modes (660-630 cm^{-1}) and C-Cl stretching (730-720 cm^{-1}). NMR experiments carried out on solutions of the tar in various deuterated solvents gave little conclusive data about the species present. However, the results do indicate the presence of CH_2Cl, CH_2O and CH_2S moieties.

Mass spectrometry has probably provided the most useful data to date. There are no significant peaks above m/z 600 and the spectra appear to confirm the presence of a series of oligomers of the form shown in Figure 10.

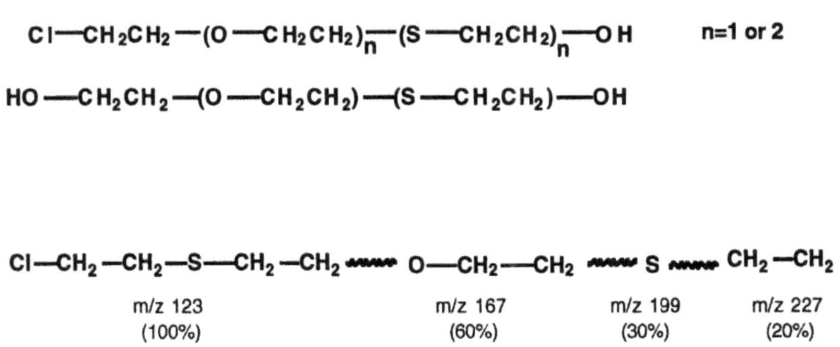

Figure 10 Oligomers in "tarry mustard", as indicated by MS

4. Silver(II) Process For The Destruction of "Tarry Mustard"

4.1 BACKGROUND

We have recently been studying the use of electrochemically generated silver(II) ions for the oxidation of CW agents in general and "tarry mustard" in particular. Patents for the process, which was developed in the UK nuclear industry, are held by AEA Technology. The process is being jointly developed by PLSD and AEA for the destruction of CW agents.

Silver(II) species can be generated electrochemically in a simple cell employing a platinum anode and a stainless steel cathode. The anode compartment contains 8 molar nitric acid and 0.5 molar silver(I) nitrate. The cathode compartment, which contains 4 molar nitric acid, is separated from the anode compartment by a Nafion membrane. Application of a 2 volt potential across the cell causes the formation of a yellow coloration in the anode compartment which is attributable to the formation of silver(II).

Silver(II) is a powerful oxidant with a redox potential of 2.0 volts versus NHE. It has been shown to oxidise a wide variety of organic substrates, but in particular has been demonstrated to oxidise "tarry mustard" through to CO_2, H_2O, Cl^- and SO_4^{2-} [7]. The chloride ion is currently deposited as AgCl, but the use of alternate oxidation systems such as Ru(VIII) should preclude this, enabling an easily regenerable system to be developed.

4.2. APPLICATION

It has been possible to show that upon addition of "tarry mustard" to a solution of silver(II), the polymeric material disperses immediately to give a white precipitate of silver chloride. Monitoring the solution for residual 'mustard' shows no detectable 'mustard' only a few seconds after dissolution, however, oxidation of the sulphur moiety to inorganic sulphate takes a further 4- 6 hours. Total analysis of the products shows a lower chlorine concentration and a higher sulphur concentration than would be theoretically possible if the starting material had been pure 'mustard'. The inference is that during the polymerisation process chlorine, or more likely hydrogen chloride, has been lost.

The silver(II) process has the potential to be made transportable. This means that it could be taken into the field to deal with small numbers of old and/or abandoned munitions. This is potentially an attractive way in which to deal with non-stockpile munitions found in small numbers in a wide variety of locations world-wide. Methods for carrying out such remediation techniques in the field have been considered, and we believe that it is possible to construct an "electrochemical incinerator" that could be deployed in the field to deal with small numbers of old munitions.

5. Conclusions

It is likely that "tarry mustard" will not have a unique composition. The nature of the material will be very dependent on the precise source and conditions of storage of the old 'mustard' involved. For this reason, destruction methods will need to be tested on a wide range of substrates if they are to be universally applicable. Total mineralisation methods would appear to be the most attractive in this respect, but with the exception of the silver(II) process, none are currently compact enough to be field deployable, but this does not preclude their use in stockpile remediation.

6. References

1. Niemann (1860)*Annalen* **113**, 288.
2. Guthrie,F. (1860) *J.Chem.Soc.* **xii,** 109
3. Stein W.H. *et.al.* (1946) Chemical Reactions of Mustard Gas and Related Compounds, *J.Org.Chem.* **11** , 664-674
4. Bartlett, P.D. and Swain, C.G. (1949) Kinetics of Hydrolysis and Displacement Reactions of Mustard Gas and Mustard Chlorohydrin, *J.Amer.Chem.Soc* **71**, 1406-1415
5. Yang, Yu-Chu *et.al.* (1987) Direct NMR Measurements of Sulphonium Chlorides Produced from the Hydrolysies of 2-Chloroethyl Sulphides, *J.Org.Chem.*, 1637-1638
6. Hage, R. *et.al.* (1994) Efficient Manganese Catalysts for Low Temperature Bleaching, *Nature* **369**, 637-639
7. Steele, D.F. *et.al.* (1990) The Low-Temperature Destruction of Organic Waste by Electrochemical Oxidation, *Trans. I ChemE.* **68(B)**, 115-121

KINETIC AND TOXICOLOGICAL PARAMETERS OF 'MUSTARD' (HD) HYDROLYSIS AND BIODEGRADATION

STEVEN P. HARVEY, TIMOTHY A. BLADES, LINDA L. SZAFRANIEC, WILLIAM T. BEAUDRY, MARK V. HALEY, THOMAS ROSSO and GERALD P. YOUNG
U.S. Army Edgewood Research, Development and Engineering Center
Aberdeen Proving Ground, MD 21010-5423

JAMES P. EARLEY
SBR Technologies, Inc.
South Bend, IN 46617

ROBERT L. IRVINE
University of Notre Dame
Notre Dame, IN 46556

1. Abstract

One technology being developed for potential use in the disposal of bis(2-chloroethyl) sulfide , otherwise known as 'mustard' or yperite or by the military code "HD", from the Aberdeen Proving Ground unitary chemical stockpile, is hydrolysis followed by biodegradation. Stockpile HD (89.2% pure) was hydrolyzed in 90° C water at a concentration of 3.8 wt %. The reaction was complete (HD undetectable at 200 ppb level) and the product was typically 90-95% 2,2'-thiodiethanol, often called thiodiglycol. The complete product was amended with 1.5 g/L NH_4Cl, 0.28 g/L KH_2PO_4, and trace mineral salts and fed to aerobic sequencing batch reactors seeded with activated sludge. The efficiency of carbon removal in the bioreactors was typically greater than 90% and the effluent was non-toxic as determined by aquatic toxicity tests.

2. Background

HD is highly insoluble in water [8] and toxic to microorganisms because of its reactivity with enzymes and other proteins [3]. For these reasons HD is a poor candidate for direct biodegradation. However, HD does react with water to form hydrolysis products that are non-chlorinated, water-soluble and significantly detoxified.

J.F. Bunnett and M. Mikołajczyk (eds.), Arsenic and Old Mustard: Chemical Problems in the Destruction of Old Arsenical and 'Mustard' Munitions, 115–122.

Consequently, these hydrolysis products are generally better candidates for biodegradation.

The mechanism of 'mustard' hydrolysis at ambient temperature has been investigated [1, 5, 6, 8, 9, 10]. It has been demonstrated that HD reacts through a series of sulfonium ion intermediates to produce thiodiglycol. However, additional ether or thioether products can also form depending on the actual conditions of the reaction. The rate of hydrolysis in water is limited by the rate of mass transfer because of the insolubility of HD in aqueous solutions. Consequently, dissolution and reaction take place simultaneously [8].

Hydrolysis dechlorinates HD providing significant detoxification. The resulting products are primarily biodegradable alcohols. The subsequent mineralization of these compounds can consequently be accomplished biologically at ambient temperature.

3. Materials and Methods

3. 1. MATERIALS

HD was obtained directly from a one-ton storage container from the U.S. Army Aberdeen Proving Ground chemical stockpile where it had been stored for approximately 40 years. It was determined to be 89.2 area % pure by gas chromatography/mass spectrometry (GC/MS). The major impurity (4.7%) was 1,2-bis(2-chloroethylthio)ethane, also known as compound Q or sesquimustard. The next in prominence was 1,2-dichloroethane (2.4%), which is probably formed by thermal decomposition of the HD dimer. $C_6H_{12}Cl_2S$ isomers comprised a total of 2.0 area % of the material. These isomers are believed to be sulfonium ion thermal decomposition products [2]. The HD was used as received for all hydrolysis and biodegradation studies.

Modified Wolin Salts Solution was prepared 100 at times usual strength as follows: 3.0 g/L nitrilotriacetic acid, 6.0 g/L. $MgSO_4 \cdot 7H_2O$, 0.5 g/L $MnSO_4$, 0.5 g/L $FeSO_4 \cdot 7H_2O$, 0.1 g/L $CaCl_2 \cdot 6H_2O$, 0.1 g/L $CoCl_2 \cdot 6H_2O$, 0.1 g/L $ZnSO_4 \cdot 7H_2O$, 0.02 g/L H_3BO_3, 0.01 g/L $CuSO_4 \cdot 5H_2O$ and 0.01 g/L $MoO_4 \cdot 2H_2O$.

The HD hydrolysis reaction profile was obtained from a 35 gallon (132.5 L) batch reaction conducted in a modified 55-gallon (208 L) stainless steel drum 30.5 inches (77.5 cm) high by 22.75 inches (57.8 cm) wide and fitted with a 5/16 inch (7.94 mm) thick polypropylene drum-liner. A nickel alloy coil circulating hot water was used to maintain the temperature at 84-86° C. Agitation (350 rpm) was provided by two impellers: a 6 inch (15.24 cm) Rushton turbine mounted one impeller diameter above the bottom of the drum to provide shearing action and a 7 inch (17.8 cm) pitched-blade axial impeller mounted at the mid-point of the impeller shaft to provide top to bottom circulation in the vessel. The impellers and shafts were coated with polyvinylidene fluoride polymers (PVDF) to prevent corrosion by the acidic hydrolysate solution. The reactor also contained four 34 inch (86.4 cm) long angle baffles consisting of 2 x 1.5 inch (5.08 x 3.81 cm) nickel alloy running vertically on the tank wall. All HD was added over a four minute period prior to the start of agitation. Sampling was done manually with a 50 ml syringe beginning 42 seconds after the start of agitation and continuing at increasing intervals up to 120 min. Samples were extracted twice, each time using 1/10 volume of chloroform. The organic layer was analyzed for HD by GC/MS and the aqueous layer was analyzed by NMR for sufonium ions and thiodiglycol.

3.2. METHODS

The MICROTOX bioassay exposes a bioluminescent marine bacterium (*Vibrio fischeri*) to a toxicant and measures the change in light output as the means of determining effects on the organisms. Light output is measured after five and 15 minutes of exposure. Assays were performed using 100% sample with two replicates of five treatment groups (1:2 serial dilution) and one control. Phenol was used as a reference toxicant. At least four dilutions were used to produce a dose response from which an EC_{50} was calculated using MICROTOX statistical software.

GC/MS analyses of HD from 35 gallon reactor samples were conducted using a Hewlett-Packard 5890 GC with 5970 mass selective detector. HD was quantified using selected ion monitoring for HD ions *m/e* 109, 111, 158, and 160. The parameters for the GC were an initial temperature of 70° C with a one minute hold followed by a 4° /min ramp to 110° C then 25° C/min ramp to a final temperature of 250° C with a five minute hold. The injector was held at 250° C with 50 ml/min split flow, which was turned on at 0.6 minute. A four point calibration curve of HD in hexane was generated prior to analysis of sample extracts. The calibrations correlation coefficent was 0.998 and the detection limit was 1 ppm.

NMR analyses were conducted using a Varian Unity Plus 400 Fourier Transform (FT) NMR spectrometer, which operates at 400 MHz for ^1H observation and at 100 MHz for ^{13}C observation. All spectra were obtained at probe temperature (22 ± 1° C) with double precision data accumulation. Samples were provided in $CDCl_3$, D_2O or water. All samples run in $CDCl_3$ were referenced to internal tetramethylsilane (TMS) using the $CHCl_3$ resonance (d^{1}H, 7.24; d^{13}C, 77.2) as a secondary reference. ^1H spectra in D_2O and water were referenced to external sodium 3-trimethylsilylpropanoate-2,2,3,3-d_4 (TSP) in D_2O. Quantitative data were obtained by digital integration of peak areas.

Survey ^1H spectra were run for product identification using a sweep width of 8000 Hz (20 ppm), a pulse width of at least 12 microseconds (30 degrees), an acquisition time of 2-4 seconds and a pulse delay of 2-4 seconds. Corresponding ^{13}C spectra were acquired using a sweep width of at least 25000 Hz (250 ppm), an acquisition time of 1.6 seconds, and a pulse delay of 2-3 seconds. Spectra were accumulated until the desired signal-to-noise ratio was achieved. Pre-saturation of the large H_2O peak was often used (standard Varian software) to minimize this resonance in the spectrum.

Spectra obtained for kinetic runs were acquired with fixed gain and integral values using the absolute intensity (AI) mode of the spectrometer. Each set of samples from a reaction was analyzed under identical operating conditions so that the integral areas from different samples from the set could be directly compared. For ^1H spectra, 16 transients were collected using a sweep width of 8000 Hz (20 ppm), an acquisition time of 2.0 seconds, a pulse width of 7 microseconds (21 degrees) and a pulse delay of 2 seconds. For ^{13}C spectra, 64 transients were collected using a sweep width of 25000 Hz, an acquisition time of 2 seconds, a pulse width of 12 microseconds (90 degrees), and a pulse delay of 2.5 seconds. ^1H data obtained in $CDCl_3$ were normalized to the residual $CHCl_3$ resonance in the solvent. Data obtained in water or D_2O were not normalized.

4. Results

4. 1. HYDROLYSIS

The essential overall HD hydrolysis reaction is described by equation (1):

$$ClCH_2CH_2SCH_2CH_2Cl + 2\ H_2O \longrightarrow HOCH_2CH_2SCH_2CH_2OH + 2\ HCl \qquad (1)$$

In actual practice, the yield of thiodiglycol is dependent on the pH of the reaction. Reactions conducted in water alone yield a relatively high percentage of thiodiglycol (90-95 area % by NMR), whereas reactions conducted in an excess of aqueous NaOH (simultaneous hydrolysis and neutralization) typically yield around 65 area % thiodiglycol, with the remaining compounds primarily ether-alcohols [4]. Thiodiglycol is biodegradable [2], consistent with the finding that the water-hydrolyzed product was more biodegradable (>90% total organic carbon removal) than the aqueous NaOH-hydrolyzed product (*ca.* 65% total organic carbon removal). Based on these observations, all hydrolysis reactions for these studies were conducted in water. NaOH was added for neutralization of the HCl subsequent to the hydrolysis reaction.

Results of experiments on the effects of temperature and HD concentration on the rate and products of the hydrolysis reaction have been described previously [4]. Briefly, the reaction proceeds relatively slowly at 30° C and is accelerated approximately 28-fold by an increase in temperature to 70° C (rates are described in relative terms due to the mass-transfer-limited nature of the reaction, which makes the rate dependent on agitation). Thiodiglycol is the main hydrolysis product; various sulfonium ions are intermediates in the reaction[9].

Figure 1 illustrates the progress of the HD hydrolysis reaction in water as followed by sampling and analysis from a 35 gallon reaction conducted at 84-86° C. Configuration, operation and sampling of the reactor are described in 3.2. METHODS above. Results in Figure 1.a show the decrease in HD concentration (plotted logarithmically against the left axis) and the increase in thiodiglycol concentration (plotted linearly against the right axis) as a function of time. The HD concentration decreases from 38,000 ppm to less than 0.2 ppm (the U.S. Army drinking water standard for HD) in less than 20 minutes, corresponding to a half-life of approximately 30 seconds. The concentrations of two sulfonium ion intermediates, CH-TG and H-2TG, are also plotted as determined by NMR analysis. The structures of these

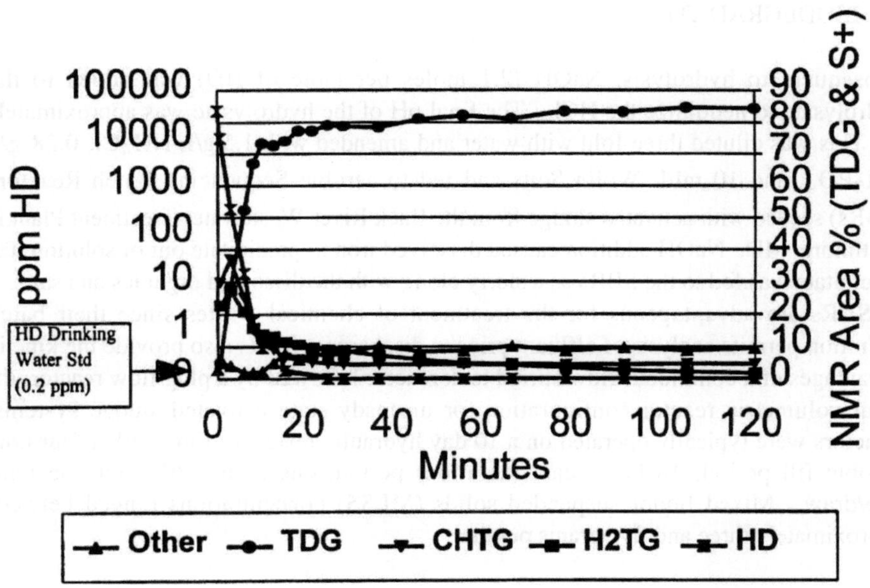

Figure 1.a. Reaction profile of HD hydrolysis reaction conducted in 35 gallon reactor.

sulfonium ions are illustrated in Figure 1.b and 1.c. The half-life of the H-2TG sulfonium ion was measured at 38 minutes under these conditions. Not shown on this graph are the concentrations of the chlorinated sulfonium ion ($(HOCH_2CH_2)_2S^+CH_2CH_2SCH_2CH_2Cl$), which was detected in the 42 second (10.6 area %) and three minute (2.0 area %) reaction samples.

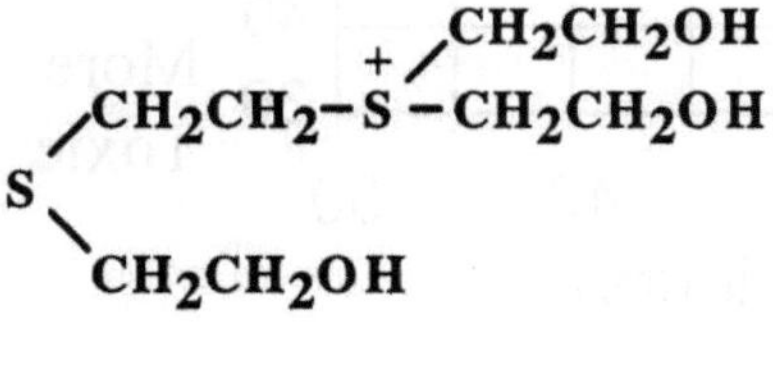

Fig. 1.b: CH-TG sulfonium ion

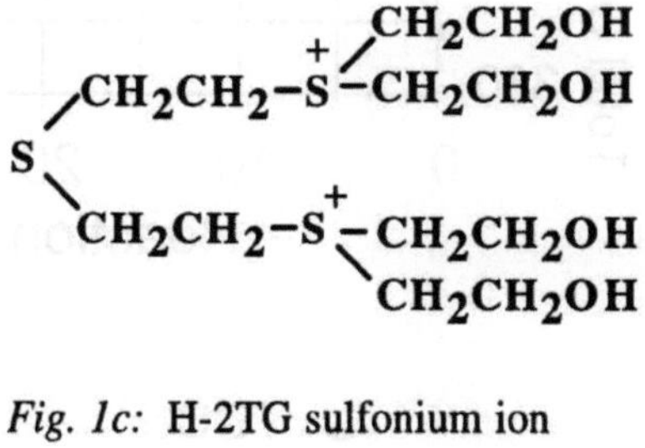

Fig. 1c: H-2TG sulfonium ion

At the end of the reaction, NMR analysis showed thiodiglycol to comprise 84.4 area % of the product, CH-TG 4.5 area %, H-2TG 1.7 area % and other unidentified compounds 9.4 area %. HD was undetectable (detection limit <150 ppb) throughout the last 100 minutes of the reaction.

4.2. BIODEGRADATION

Subsequent to hydrolysis, NaOH (2.1 moles per mole of HD) was added to the hydrolysate to neutralize the HCl. The final pH of the hydrolysate was approximately 11. This was diluted three-fold with water and amended with 1.5 g/L NH_4Cl, 0.28 g/L KH_2PO_4, and 10 ml/L Wolin Salts and fed to aerobic Sequencing Batch Reactors (SBRs) seeded with activated sludge from the Back River Wastewater Treatment Plant in Baltimore, MD. NaOH addition caused dissolved iron to precipitate out of solution; the precipitate was fed to the SBRs as a slurry along with the dissolved organics and salts.

SBRs are advantageous for the treatment of chemical wastes since their batch operation permits analysis of effluent prior to discharge. SBRs also provide the kinetic advantages of a continuous flow stirred tank reactor followed by a plug flow reactor; the ideal volumetric reactor configuration for unsteady state activated sludge systems. Reactors were typically operated on a 10 day hydraulic residence time with a four hour aerobic fill period, 18 hour react (aeration) period, one hour settle and one hour idle/draw. Mixed liquor suspended solids (MLSS) concentrations ranged between approximately three and five grams per liter.

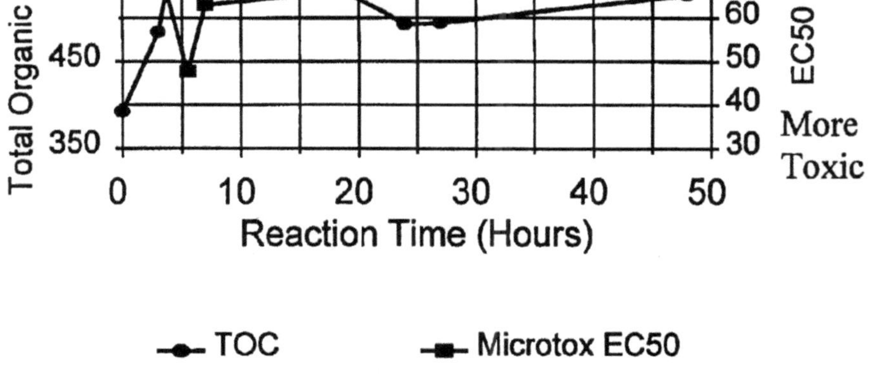

Figure 2. TOC and MICROTOX toxicity profiles versus reaction time for HD hydrolysate biodegradation process.

4.2.1. Bioreactor Effluent Analysis

The bioreactor effluent analysis showed an average removal efficiency of 99.6% for thiodiglycol and 90.6% for total organic carbon (TOC). Figure 2 shows the TOC profile and the results of MICROTOX luminescent marine bacteria toxicity testing versus reaction (aeration) time. The MICROTOX data are presented as EC_{50} values. EC_{50} is defined as the effective concentration of material that decreases luminescence (viability) of the test organisms by 50%. The toxicity initially decreases and the TOC increases during the fill period when feed is being added to the reactor. As the fill is completed and biodegradation continues, the TOC drops and the toxicity decreases (i.e. the EC_{50} value increases). Although organic carbon removal is essentially complete within a 22 hour react period (a 10 day hydraulic residence time as the reactors are typically operated), the decrease in toxicity of the bioreactor effluent continues for approximately twice that long (up to 48 hours react which would be equivalent to a 20 day hydraulic residence time).

Although the exact cause of the discontinuity between TOC removal and toxicity reduction is not known, a simple explanation would be that one or more compounds of relatively minor abundance account for a significant portion of the detectable toxicity and that these compounds are biodegraded at a rate slow relative to that of the major portion of the organic carbon. Not shown in Figure 2 is the toxicity of the starting feed material which had an EC_{50} of approximately 2%. The final effluent, biodegraded for the equivalent of a 20 day hydraulic residence time (48 hours reaction with a 10 day hydraulic residence time) at the above conditions, was completely non-toxic to luminescent marine bacteria.

5. Conclusions

HD, at a 3.8% loading in water, in a well-mixed reactor at 84-86° C, reacts with a half-life of approximately 30 seconds to a concentration below 200 ppb within 20 minutes. The half-life of the H-2TG sulfonium ion under these same conditions is 38 minutes. The hydrolysate, as analyzed by NMR, is comprised primarily of thiodiglycol with small amounts of H-2TG and CH-TG sulfonium ions.

When the hydrolysate was diluted with three volumes of water, amended with 1.5 g/L NH_4Cl, 0.28 g/L KH_2PO_4, and mineral salts and fed to aerobic SBRs seeded with activated sludge (MLSS = three to five g/L), greater than 99% of the thiodiglycol and greater than 90% of the total organic carbon were removed (converted to CO_2 or biomass). MICROTOX studies demonstrated complete detoxification with a react time of 48 hours although TOC removal was essentially complete after 24 hours react time.

6. Acknowledgments

The authors offer their sincere gratitude to Craig M. Frost of SBR Technologies, Inc., for his dedicated efforts operating the bioreactors, to Dennis K. Rohrbaugh of ERDEC, for his GC/MS analysis of the stockpile HD, and to Harry G. Taylor, and George Roberts of ERDEC for their expert assistance on the operation of the 35 gallon chemical reactor. This research was funded by the U.S.Army Program Manager for Chemical Demilitarization.

7. References

1. Bartlett, P.D. and Swain, C.G. (1949) Kinetics of Hydrolysis and Displacement Reactions of β,β'-Dichlorodiethyl Sulfide (Mustard Gas) and of β-chloro-β'-Hydroxydiethyl Sulfide (Mustard Chlorohydrin), *J. Am. Chem. Soc.* **71**, 1406-1415.
2. Beaudry, W.T., Bossle, P.C., Harvey, S.P., Kolakowski, J.E., Procell, L.R., Rohrbaugh, D.K., Sorrick, D.C., Stroup, A.N., Szafraniec, L.L. and Yang, Y.C. (1994) Neutralization of Munitions-Grade HD to Biodegradable Components. Proceedings, 19th Army Science Conference, Assuring the Competitive Edge. Army Science Board, eds. Vol. 4, pp. 1497-1504. Washington D.C.
3. Bush, V. (1946) *Summary of Technical Report of Division 9, National Defense Research Committee*. Washington, D.C.: Columbia University Press.
4. Harvey, S.P., Szafraniec, L.L., Beaudry, W.T., Earley, J.P. and Irvine, R.L. Neutralization and Biodegradation of Sulfur Mustard, in S.K. Sikdar and R.L. Irvine (eds.), *Bioremediation: Principles and Practice Created*, Vol. II: Biodegradation Technology Developments, Technomic Publishing Inc. In press.
5. Helfrich, O.B. and Reid, E.E. (1920) Reactions and Derivatives of β,β'-Dichloro-ethyl Sulfide, *J. Am. Chem. Soc.* **42**, 1208-1232.
6. McManus, S.P., Neamati-Mazrach, N., Hovanes, B.A., Paley, M.S. and Harris, J.M. (1985) Hydrolysis of Mustard Derivatives. Failure of the Raber-Harris probe in predicting nucleophilic assistance, *J. Am. Chem. Soc.* **107**, 3393-3395.
7. Rohrbaugh, D.K., Yang, Y.-C., and Ward, J.R. (1989) The characterization of sulfonium chlorides by gas chromatography/mass spectrometry and the degradation of 2-chloroethyl sulfide derivatives, *Phosphorus, Sulfur and Silicon* **44**, 17-25.
8. Yang, Y.-C., Szafraniec, L.L., Beaudry, W.T. and Ward, J.R. (1987) Direct NMR Measurements of Sulfonium Chlorides Produced from the Hydrolyses of 2-Chloroethyl Sulfides, *J. Org. Chem.* **52**, 1637-1638.
9. Yang, Y-C., Szafraniec, L.L., Beaudry, W.T. and Ward, J.R. (1988) Kinetics and Mechanism of the Hydrolysis of 2-Chloroethyl Sulfides, *J. Org. Chem.* **53**, 3293-3297.
10. Yang, Y.C., Ward, J.R., Wilson, R.B., Burrows, W. and Winterle, J.S. (1987) On the Activation Energy for the Hydrolysis of Bis-(2-chloroethyl) Sulfide. 11, *Thermochim. Acta.* **114**, 313-317.

ARSENIC IN THE ENVIRONMENT

W.R. CULLEN
Department of Chemistry, University of British Columbia
2036 Mail Mall, Vancouver, BC CANADA V6T 1Z1

1. Abundance and Occurrence

Arsenic is a member of Group 15 of the Periodic Table. This used to be known as Group VB (or Group VA) depending on whose textbook was being used. The elements in the Group are as follow: N, P, As, Sb, Bi, of atomic numbers, respectively, 7, 15, 33, 51, and 83. The metallic character increases with increasing atomic number. Nitrogen and phosphorus are essential for life and much is known about their biochemistry. Arsenic may be essential for some forms of life, but nothing is known about how this element could be involved in life processes.

An important difference between P and As is the ease of interconversion between the (V) and (III) oxidation states. In nature almost all phosphorus compounds are in the (V) oxidation state (or 5 valent state). Arsenic compounds are usually in either the (V) or (III) state. Arsenic metal has (0) oxidation state assigned to it. To put some numbers on the relative ease of oxidation/reduction of arsenic over phosphorus, the following half cell potentials are instructive. Thus, whereas arsenate is easily reduced to arsenite in acid solution, the reverse is true, easy oxidation of arsenite to arsenate, in base.

			E°(V)
in acid	$H_3PO_4 + 2\,H^+ + 2e$	$= H_3PO_3 + H_2O$	-0.267
	$H_3AsO_4 + 2\,H^+ + 2e$	$= H_3AsO_3 + H_2O$	0.559
in base	$PO_4^{3-} + H_2O + 2e$	$= PO_3^{3-} + 2\,OH^-$	-1.12
	$AsO_4^{3-} + H_2O + 2e$	$= AsO_3^{3-} + 2\,OH^-$	-0.67

Arsenic is found everywhere, including outer space (1). In natural waters (fresh and salt), the concentration of arsenic is normally in the parts per billion, ppb, range; in the outer earth's crust the concentration is in the parts per million, ppm, range, although local highs can be found, *e.g.* up to 6,000 ppm in some pyrite (FeS_2) samples (2). Some natural sources of arsenic are listed in Table 1.

J.F. Bunnett and M. Mikołajczyk (eds.), Arsenic and Old Mustard: Chemical Problems in the Destruction of Old Arsenical and 'Mustard' Munitions, 123–134.

TABLE 1: Natural Arsenic-Containing Minerals

Endogenous[a]
realgar, As_4S_4; orpiment As_4S_6;
arsenical pyrites, mispickel, FeAsS
niccolite NiAs

Supergenous[b]
arsenolite, claudetite, As_4O_6
adamite $Zn_2(AsO_4)(OH)$ ("zinc arsenate")
erythrite $Co_3(AsO_4)_2 \cdot 8H_2O$ (cobalt arsenate)

[a]Formed below the surface [b]Formed mainly via oxidation of endogenous species.

Weathering results in the release of arsenic in soluble forms and it is estimated (3) that ~45 x 10^6 kg/year enter the ocean. Atmospheric contributions comprise very roughly 18 x 10^8 kg/year from volcanic activity, and 26 x 10^6 Kg/year from soils and sediments as volatile arsines (4). Human activities result in introduction of (a) up to 26 x 10^6 kg/year to the atmosphere (5), mainly from nonferrous metal production; (b) about 100 x 10^6 kg/year to land from mine tailings, smelter slags and water, coal fly ash and bottom ash, and commercial product wastage, etc., and (c) about 40 x 10^6 kg/year into aquatic systems from domestic waste water, electric generating plants, and ferrous smelters, *etc.* (6).

2. Affinity of Arsenic for Sulfur

The affinity of arsenic for sulfur is seen in the list of common minerals, Table 1. Readers may recall that arsenic is usually precipitated as the sulfide in qualitative analysis schemes. Many proteins contain -SH groups and these readily react with arsenic compounds in both oxidation states. As(V) is reduced to As(III), which then binds to the protein and, as a consequence, interferes with its normal function.

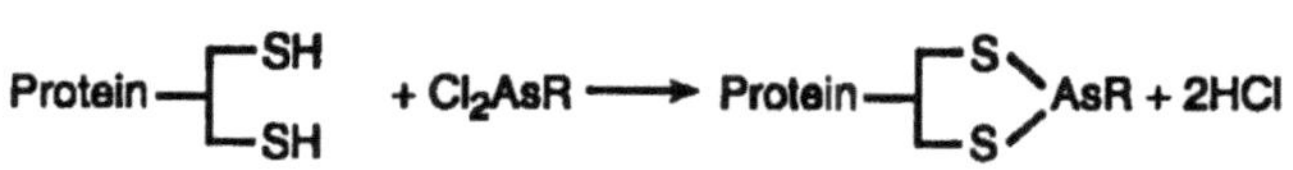

(1)

One way of restoring function is to load up the system with an agent that will preferentially bind the arsenic. Such an agent is BAL (British Anti Lewisite), which was developed as an antidote to Lewisite (R = -CH = CHCl)

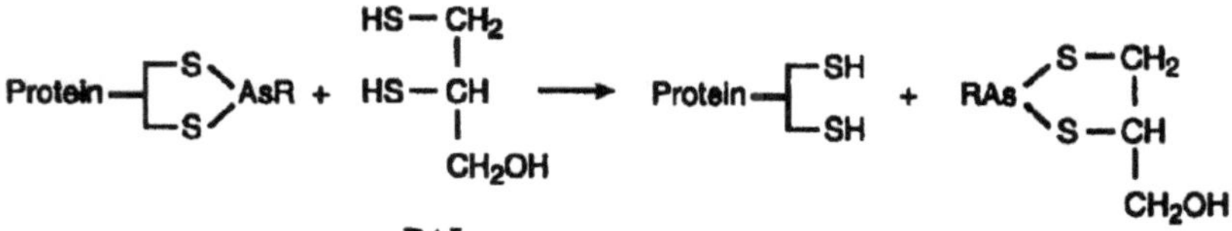

(2)

BAL

3. Biological Methylation of Arsenic

In European homes in the late 19th century it became obvious that a toxic gas with a garlic-like odor could be produced bythe mildewing of colored wallpaper. Some green pigments used in the wallpaper at that time were arsenic compounds, *e.g.* Scheele's green, copper arsenate; the gas, known as Gosio gas, was later [1932] identified as trimethylarsine, $(CH_3)_3As$, by Challenger and his co-workers (7): A modern version of his experiment is shown in Figure 1 (8).

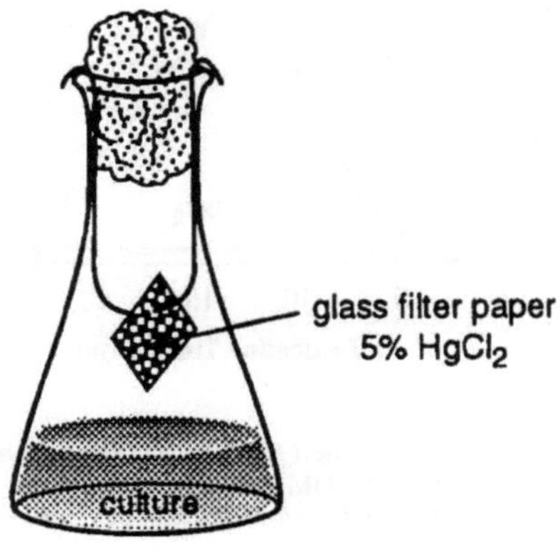

Figure 1: The use of a prepared filter paper to chemofocus volatile metabolites.

The arsenic compound is added to the culture. The gas is evolved and trapped on the filter paper because it reacts with the $HgCl_2$ to form white crystals of the adduct: $(CH_3)_3As + HgCl_2 \longrightarrow (CH_3)_3As{\cdot}HgCl_2$. The filter paper can be analyzed by using mass spectrometry directly, or by heating to liberate the arsine, which is then analyzed by using other methods such as gas chromatography with or without element-specific detection. If radio-labels are involved, the paper can be counted directly to give a measure of the biotransformation. Many molds and fungi have the ability to methylate arsenic compounds under aerobic conditions. For example, *Apiotricum humicola*, formerly known as *Candida humicola*, methylates arsenate to Me_3As (9).

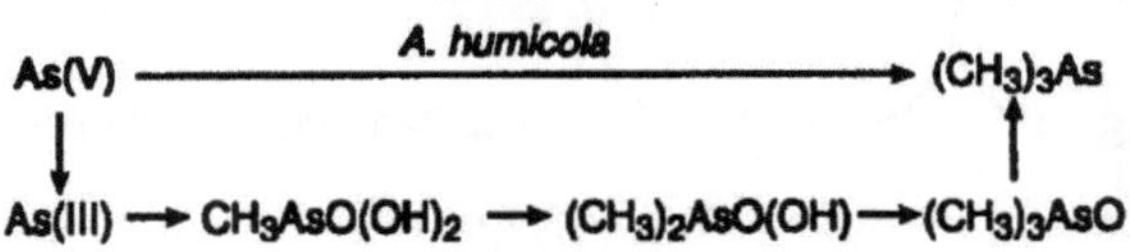

(3)

At low arsenic concentrations the arsine is not formed although all the intermediates from As(III) to $(CH_3)_3AsO$, shown along the bottom pathway of equation 3, can be detected in the culture (10). A time course study of this process is shown in Figure 2.

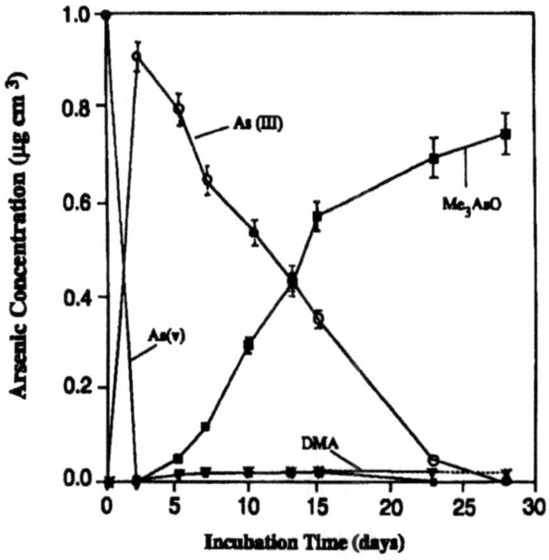

Figure 2: The biotransformation of arsenate (As(V), 1 ppm) by *Apiotrichum humicola* to first arsenite As(III) and finally $(CH_3)_3AsO$. DMA = $(CH_3)_2AsO(OH)$. The concentration of the species was monitored by using hydride generation methodology: see Figure 4.

Bacteria, anaerobic and aerobic, have the same ability to methylate arsenic. The volatile products can be fully methylated, Me_3As, but they can also be arsines with As-H bonds, *e.g.* $MeAsH_2$. For example, starting with As(V), *Pseudomonas* sp affords As(III), CH_3AsH_2, $(CH_3)_2AsH$, and $(CH_3)_3As$ and *Flavobacterium* sp affords As(III) and $(CH_3)_2AsH$ (11).

The mechanism of the biomethylation of arsenic is generally believed to involve a series of reduction/oxidation steps, one cycle of which is shown in Figure 3 (11).

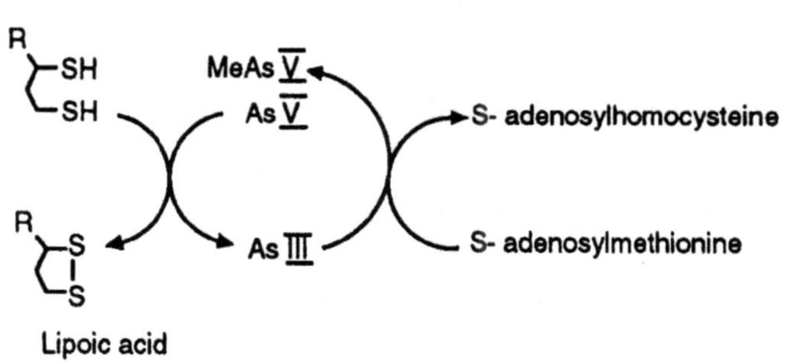

Figure 3: One step of the reduction/oxidation cycle that results in the biological methylation of arsenic.

Reduction of As(V) to As(III), probably by a thiol such as lipoic acid or glutathione, is followed by oxidative transfer of a methyl group (CH_3^+) from S–adenosylmethionine (SAM) to yield the $CH_3As(V)$ product. The cycle is repeated to afford a $(CH_3)_2As(V)$ derivative and then $(CH_3)_3AsO$ (bottom line of Eq. 3), which may or may not be reduced to volatile $(CH_3)_3As$ (see also Fig. 6). In the experiment shown in Figure 2, reduction of $(CH_3)_3AsO$ does not occur if the arsenic concentration is less than 1 ppm (10).

4. Toxicological Properties

The gas arsine AsH_3 (m.p. -116°C; b.p. –55°C) is very toxic, is easily produced from inorganic arsenic compounds, and can be the cause of industrial accidents. The oxide As_4O_6 (As_2O_3), white arsenic, is the chemical that the general population associates with the word arsenic. For many years it was the substance of choice for homicidal use and a fatal dose for humans is in the 70-180 mg range. The general sequence for the toxicity of arsenicals is as follows (12): H_3As, R_3As > inorganic As(III), arsenite > organic As(III) *e.g.* $(RAsO)_n$ > inorganic As(V), arsenate > organic As(V) *e.g.* $R_2AsO(OH)$ > R_4As^+ > elemental arsenic.

Not much is known about the toxicity of the organoarsenicals found in the environment, although in general, organic substituents reduce the toxicity of arsenic species substantially. Some data for mice illustrate this effect (13) (compound, LD_{50} mg kg^{-1}) As_4O_6, 35; $CH_3AsO(OH)ONa$, 1,800; $(CH_3)_2AsO(ONa)$, 1,200; $(CH_3)_4AsCl$, 580; $(CH_3)_3As^+CH_2COO^-$, > 10,000. The arsenic(III) species $(RAsO)_n$ (R = CH_3, C_6H_5, -CH = CHCl) are more toxic than arsenite in an assay that involves their ability to block gluconeogenesis in rat kidney tubules and canine kidney cells (14) . However, with very few exceptions, it is very difficult to clarify the effect of arsenicals on humans by using animal studies. For example, arsenicals can undoubtedly cause cancer of the skin in man, but it has never been possible to demonstrate such an effect on any other animal. In spite of this, the International Agency for Research on Cancer has evaluated arsenic as a Group I carcinogen, i.e. is carcinogenic to humans (15). The classification is based mainly on studies involving inorganic arsenic species. What is known about carcinogencity and organoarsenicals? It seems that dimethylarsinic acid, $(CH_3)_2AsO(OH)$, Agent Blue (Table 2), may not be as benign as would be expected from toxicity data such as that given above for mice. The arsenical seems to enhance the tumor development in the urinary tract, liver, and thyroid glands of rats (16).

The Canadian Environmental Protection Act recently required an assessment of arsenic and its compounds to determine whether they are "toxic" as defined in the Act, *i.e.* (a) has ... an immediate or long-term harmful effect ...; (b) changes the environment on which human life depends; (c) constitutes a danger in Canada to human life or health... The conclusion was that inorganic arsenic compounds are toxic on both (a) and (c) criteria, but there are not enough data to justify a conclusion regarding organic derivatives.

Arsenic compounds that have been used in warfare or in crowd control, or whose use has been seriously contemplated for these purposes, are listed in Table 2 (17). HL is a mixture of Lewisite, L, and "mustard" gas, HD.

Table 2: Arsenicals in War

Blister Agents	MD[a]	CH_3AsCl_2	
	PD	$C_6H_5AsCl_2$	
	ED	$C_2H_5AsCl_2$	
	L	$ClCH{=}CHAsCl_2$	
	HL	$ClCH{=}CHAsCl_2$	L
		$(ClCH_2CH_2)_2S$	HD
Blood Agents	SA	AsH_3	
Vomiting Agents	DA	$(C_6H_5)_2AsCl$	
	DC	$(C_6H_5)_2AsCN$	
	DM	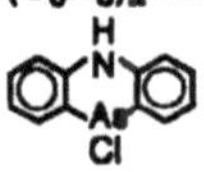	
Herbicides	Agent blue	$(CH_3)_2AsO(OH)$	

[a] STANAG Code (Standard NATO Agreement)

In addition, a recent list of chemical warfare agents (18) includes two more 2-chlorovinyl derivatives $(ClCH{=}CH)_3As$, L-3, and $(ClCH{=}CH)_2AsCl$, L-2: L-1 is the description used for the architypical Lewisite, L in the Table. Apart from the herbicide $(CH_3)_2AsO(OH)$, use of these compounds was not conceived because of the toxic properties associated with "arsenic" poisoning.

5. Arsenic Species in the Terrestrial Environment

Arsenic species that have been identified in the terrestrial environment are listed in Table 3. Apart from the inorganic species, which predominate in all environmental compartments, they are mainly methylarsenicals and are presumably formed via the same biological process outlined above. The formulations given for the methylarsenic(III) species are probably not correct because compounds of this type are unknown. It is likely that the species are actually thiols $CH_3As(SR)_2$ and $(CH_3)_2AsSR$ (19). The reason for the uncertainty is that the analytical technique commonly used to determine arsenic species is hydride generation (HG) followed by some form of separation and detection, *e.g.* gas chromatography (GC) and atomic absorption (AA) spectrometry; hence HG/GC/AA.

Table 3: Arsenic In The Terrestrial Environment

As (III)	As (V)
$As(OH)_3$	H_3AsO_4
"$CH_3As(OH)_2$" "$(CH_3)_2AsOH$"	$CH_3AsO(OH)_2$ $(CH_3)_2AsO(OH)$ $(CH_3)_3AsO$ $(CH_3)_4As^+$
AsH_3 CH_3AsH_2, $(CH_3)_2AsH$ $(CH_3)_3As$	$(CH_3)_3As^+CH_2COO^-$ $(CH_3)_3As^+CH_2CH_2OH$
$As(C_2H_5)_3$ $CH_3As(C_2H_5)_2$ $(CH_3)_2AsC_2H_5$	$C_6H_5AsO(OH)_2$

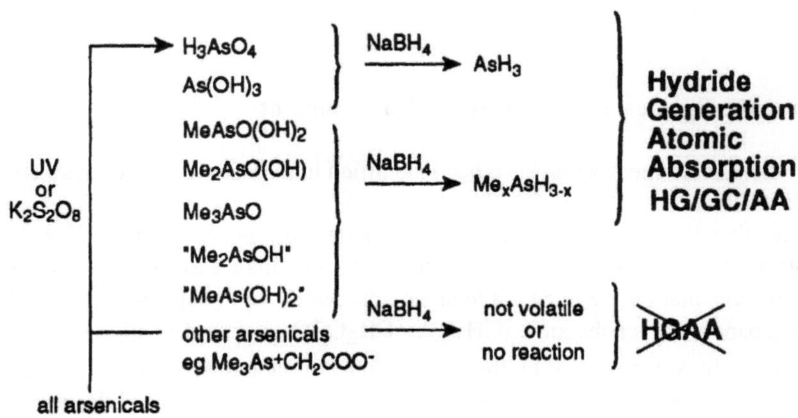

Figure 4: Speciation of arsenicals by using hydride generation methodology.

Figure 4. Information about groups other than organic is lost in this derivatization procedure. It should also be noted that this is a technique designed to "find" the arsenic compounds belonging to a specific class of compounds. Other arsenicals, which do not form volatile hydrides on treatment with borohydride, are "hidden" to this methodology.

The ethylarsines shown in Table 3 were discovered in natural gas (20); it is conceivable that they were formed by the addition of ethylene to an arsine, *e.g.*

$$(CH_3)_2AsH + CH_2{=}CH_2 \rightarrow (CH_3)_2AsCH_2CH_3 \quad (4)$$

The phenylarsonic acid was identified from shale oil (21) As for the ethylarsenicals, it is hard to conceive of a biological process for the formation of this arylarsonic acid. Perhaps a direct reaction between inorganic arsenic species and arenes can occur under geological conditions.

Little is known about the arsenic species present in plants, although arsenic concentrations can get very high. This observation has been used in geochemical prospecting because arsenic is often associated with gold (22). Recent work with mushrooms has revealed the surprising presence of arsenobetaine, $(CH_3)_3As^+CH_2COO^-$, arsenocholine $(CH_3)_3As^+CH_2CH_2OH$, and tetramethylarsonium ion $(CH_3)_4As^+$(23,24). Arsenobetaine is the major arsenic species found in marine animals (*q.v.*) and was believed to be confined to the marine environment. Likewise for arsenocholine, although this is not a major marine species. There is no obvious biological route to these compounds. Very recently Irgolic and his co-workers (24) have found arsenobetaine in the urine of a male subject who was exposed only to inorganic arsenicals and/or trimethylarsine, *i.e.*, he did not come by the arsenobetaine through eating seafood. Further information on the terrestrial environment is given in the chapter by S. Maeda in this volume.

6. Arsenic Species in the Marine Environment

Most of the arsenic species that have been identified in the marine environment are listed in Table 4 (25).

In seawater the inorganic species predominate, but, as with terrestrial water, the concentration of arsenic(III) is usually much greater than expected on the basis of equilibrium calculations. Biological reduction processes are undoubtedly responsible for this discrepancy. Arsenobetaine, $(CH_3)_3As^+CH_2COO^-$, is found in all marine animals: for example in shrimp muscle the total arsenic concentration is ~ 10 ppm and arsenobetaine is the major species found (26). Fortunately this arsenical does not appear to be harmful to humans and is excreted, rapidly and unchanged, in urine (27). The tetramethylarsoinium ion is found mainly in clams: *e.g.* ~ 50% of the total arsenic (~ 2 ppm) in the Manila clam (28).

The remaining compounds in Table 4 are mainly ribose derivatives that are encountered by humans mainly in macroalga, and as such are a part of the diet of a large segment of the world's population. As an example, the arsenic content of *Nori*, a commonly found seaweed product, is ~ 10 ppm. Essentially all this arsenic is present as one to four arseno-sugar derivatives, depending on the brand (29). Another product known as *Hijiki* has its arsenic content (~ 10 ppm) almost equally split between arseno-sugars and arsenate, which make this product more of a hazard to humans (30). The arseno-sugars are metabolized in humans and the principal metabolite found in the urine is the more toxic $(CH_3)_2AsO(OH)$ (27).

TABLE 4. Arsenic in the Marine Environment.

H_3AsO_4, $As(OH)_3$, $CH_3AsO(OH)_2$, $(CH_3)_2AsO(OH)$, $(CH_3)_3AsO$, $(CH_3)_3As^+CH_2COO^-$, $(CH_3)_3As^+CH_2CH_2OH$, $(CH_3)_4As^+$

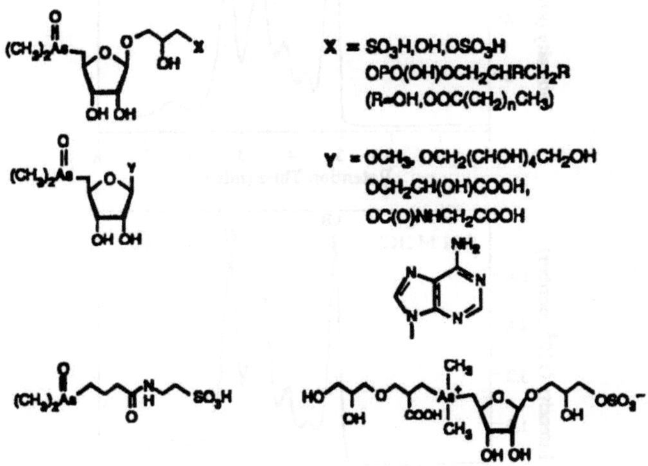

Sophisticated analytical techniques are needed for the analysis of these surprisingly complex arsenicals. The current method of choice involves separation of the water-soluble compounds by using High Performance Liquid Chromatography connected to an Inductively Coupled Plasma Mass Spectrometer (HPLC/ICP/MS) for element-specific detection. An example of such a chromatogram is shown in Figure 5 (31). It should be mentioned that much of the earlier work found only arsenobetaine in marine animal samples because of the analytical methodology that was used. A case in point is oyster tissue: the arseno-sugar seen in Figure 5 would not have been detected in earlier work.

The outcome of donating the three groups on the sulfur atom of S-adenosyl-methionine, SAM, is shown in Figure 6. It is easy to see how the arseno-sugars could arise from such a process. Indeed the compound formed by transfer of the adenosyl moiety has been isolated from the kidney of the giant clam: Table 4 (32). With regard to the formation of arsenobetaine and the tetramethylarsonium ion, the latter is easily envisaged as the product of methylating $(CH_3)_3As$. There is evidence to suggest that arsenobetaine is a breakdown product of arseno-sugars (25); however, the discovery of arsenobetaine in terrestrial mushrooms with no known arseno-sugars around to act as precursors casts doubt at least on the universality of the argument.

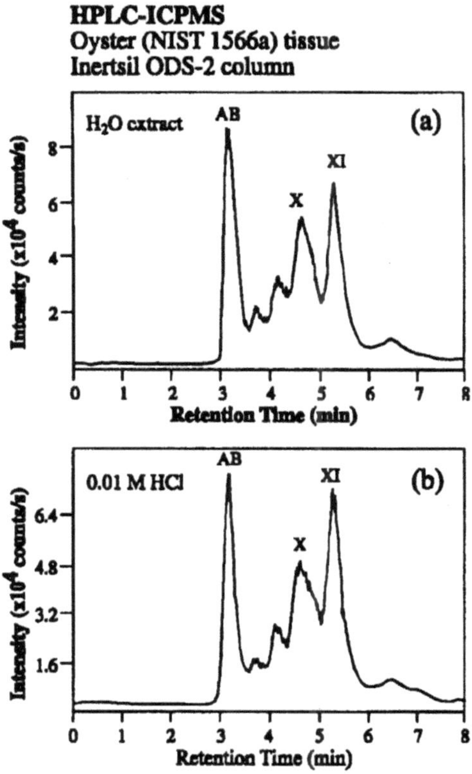

Figure 5:(a) HPLC-ICPMS trace of an aqueous extract of a sample of oyster tissue, a readily available standard reference material. AB†= arseno betaine, X and X1 are known arsenosugars. (b) The lower trace shows that the sugars are stable to 0.01†M HCl, the acidity of the stomach.

7. Biological Demethylation

Demethylation has been observed in a number of laboratory experiments. For example, sediment bacteria degrade tetramethylarsonium mainly to $(CH_3)_3AsO$ and $(CH_3)_2AsO(OH)$ (33). Similar experiments with arsenobetaine released $(CH_3)_2AsO(OH)$ (34). When seaweed containing some of the arseno-sugars shown in Table 4 is allowed to "rot" in the presence of sediment and seawater, the principal arsenic-containing product is $(CH_3)_2As(O)CH_2CH_2OH$, demonstrating biological cleavage of the ribose ring (35). This particular metabolite has been identified only in the laboratory. Nevertheless, it does appear that once arsenic is mobilized by one means or another, there are biological cycles in place that ensure that it remains in circulation.

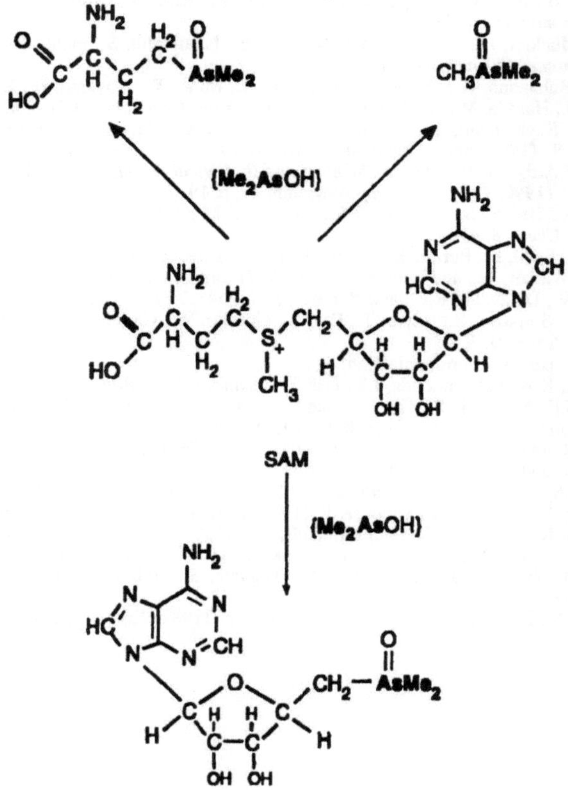

Figure 6: SAM, R[1]R[2]R[3]S+, appears to be able to transfer all three groups, R[1], R[2], and R[3] to arsenic (25).

8. References

†1. Cardelli, J.A. (1994) *Science* **265**, 204.
†2. Boyle, R.W., and Jonasson, I.R. (1973) *J. Geochem. Explor.* **2**, 251.
3. Ferguson, J.F. and Gavis, J. (1972) *Water Res.* **6**, 1259.
4. Chilvers, D.C. and Petersen, P.J. (1987) "Global Cycling of Arsenic" in T.C. Hutchinson and K.M. Meema (eds.), *Lead, Mercury, Cadmium and Arsenic in the Environment*, Scope 31, Wiley, NY, p. 279.
5. Nriagu, J.O. and Pacyna, J.M. (1988) *Nature* **333**, 134.
6. Nriagu, J.O. and Azcue, J.M. (1990) "Food Contamination from Environmental Sources" in T.O. Nriagu and M.S. Simmons (eds.) Food Contamination with Arsenic in the Environment, *Advances in Environmental Science and Technology*, Vol. 23, Wiley, New York, p. 121.
7. Challenger, F. (1945) *Chem. Rev.* **36**, 315.
8. Cullen, W.R., Erdman, A.E., McBride, B.C., and Pickett, A.W. (1983) *J. Microbiol. Methods* **1**, 297.
9. Cullen, W.R., Froese, C.L., Lui, A., McBride, B.C., Patmore, D.J., and Reimer, M. (1977) *J. Organomet. Chem.* **139**, 61.
10. Cullen, W.R., Li, H., Hewitt, G., Reimer, K.J., and Zalunardo, N. (1994) *Appl. Organomet. Chem.* **8**, 303.
11. Cullen, W.R. and Reimer, K.J. (1989) *Chem. Rev.* **89**, 713.
12. Penrose, W.R. (1974) *CRC Crit. Rev. Environ. Control*, 465.

13. Kaise, T., Yamauchi, H., Horiguchi, Y., Tani, T., Watanabe, S., Hirayama, T., and Fukui, S. (1989) *Appl. Organomet. Chem.* **3**, 273.
14. Liebl, B., Müekter, H., Nguyen, Ph-T., Doklea, E., Islambouli, S., Fichtl, B., and Forth, W. (1995) *Appl. Organomet Chem.* **9**, 531.
15. Tsuda, T., Babazono, A., Ogawa, T., Hamada, H., Mino, Y., Aoyama, H., Kurumatani, N., Nagira, T., Hotta, N., Harada, M., and Inomata, S. (1992) *Appl. Organomet. Chem.* **6**, 309.
16. Yamamoto, S., Konishi, Y., Murai, T., Shibada, M.A., Matsuda, T., Kuroda, K., Endo, G., and Fukushima, S. (1994) *Appl. Organomet. Chem.* **8**, 197.
17. Crompton, J.A.F. (1987) *Military Chemical and Biological Agents*, Telford Press, Caldwell, NJ.
18. Ember, L.R. (1996) *Chem. and Eng. News* Jan. 29, p. 19.
19. Bright, D.A., Brock, S., Cullen, W.R., Hewitt, G.M., Jafaar, J., and Reimer, K.J. (1994) *Appl. Organomet. Chem.* **8**, 415.
20. Irgolic, K.J., Spall, D., Puri, B.K., Llger, D., and Zingaro, R. (1991) *Appl. Organomet. Chem.* **5**, 117.
21. Fish, R.H., Walker, W., and Tannous, R.S. (1987) *Energy Fuels* **1**, 243.
22. Warren, H.V., Delavault, R.E., and Barakso, J. (1964) *J. Econ. Geol.* **59**, 1381.
23. Byrne, A.R., Slejkovc, Z., Stijve, T., Fay, L., Gössler, W., Gailer, J., and Irgolic, K.J. (1995) *Appl. Organomet. Chem.* **9**, 305.
24. Irgolic, K.J., personal communication.
25. Francesconi, K.A. and Edmonds, J.S. (1993) *Oceanogr. Mar. Biol. Ann. Rev.* **31**, 111.
26. Lawrence, J.F., Michalik, P., Tam, G., and Conacher, H.B.S. (1986) *J. Agric. Food Chem.* **34**, 315.
27. Le, X.C., Cullen, W.R., and Reimer, K.J. (1994) *Clinical Chem.* **40**, 617.
28. Cullen, W.R. and Dodd, M. (1989) *Appl. Organomet. Chem.* **3**, 79.
29. Cullen, W.R. and Lai, V., unpublished results.
30. Edmonds, J.S., Morita, M., and Shibata, Y. (1987) *J. Chem. Soc., Perkin Trans. 1*, 577.
31. Le, X.C., Cullen, W.R., and Reimer, K.J. (1994) *Environ. Sci. Technol.* **13**, 1598.
32. Francesconi, K.A., Stick, R.V., and Edmonds, J.S. (1991) *J. Chem. Soc., Chem. Commun.* 928.
33. Hanaoka, K., Araki, N., Tagawa, S., and Kaise, T. (1994) *Appl. Organomet. Chem.* **8**, 201.
34. Hanaoka, K., Hasegawa, S., Kawabe, N., Tagawa, S., and Kaise, T. (1990) *Appl. Organomet. Chem.* **4**, 239.
35. Edmonds, J.S., Francesconi, K.A., and Hanson, T.A. (1982) *Experientia* **38**, 643

BIOTRANSFORMATION OF ARSENIC IN FRESHWATER ORGANISMS

SHIGERU MAEDA
Department of Applied Chemistry & Chemical Engineering, Kagoshima University
1-21-40, Korimoto, Kagoshima, 890, Japan

1. Introduction

There seems to be a significant difference in level and chemical forms of arsenic between terrestrial and marine organisms. Terrestrial organisms rarely contain more than 1 μg As/g (dry weight), whereas marine organisms contain from several μg As/g to more than 100 μg As/g (Lunde[21]). The source of arsenic in marine ecosystem is, of course, inorganic arsenate [abbreviated as As(V)] and arsenite [abbreviated as As(III)] dissolved in seawater and the concentration is generally constant at a few μg As/dm^3 in all seas. On the other hand, arsenic concentrations in unpolluted rivers, ponds, and lakes are of ppb order, and the concentrations are greatly affected by circumstances.

Chemical structures of the arsenic compounds in marine organisms have been confirmed in many cases, but very few chemical species of arsenic in freshwater organisms have been found. Arsenic transformation via the freshwater food chain has rarely been reported. This chapter focuses on the toxicity of arsenicals and the biotransformation of arsenic in the freshwater organisms.

2. Background Literature

A number of books and review articles treat arsenic species and their transformation in the environment. Particular mention should be made of one book, *Arsenic*, by a U.S. National Research Council Committee on Medical and Biologic Effects of Environmental Pollutants [1]. The Third Spurenelement-Symposium resulted in a useful volume, *Arsen*[2], as did a symposium sponsored by the Chemical Manufacturers Association and the National Bureau of Standards, *Arsenic: Industrial, biomedical, environmental perspectives* [3], and the First Arsenic Symposium sponsored by the

J.F. Bunnett and M. Mikołajczyk (eds.), Arsenic and Old Mustard: Chemical Problems in the Destruction of Old Arsenical and 'Mustard' Munitions, 135–148.

Japanese Arsenic Scientists Society (JASS) , *Arsenic: Chemistry, metabolism and toxicity* [4]. The JASS symposia resulted in successive volumes as special issues of *Applied Organometallic Chemistry*, entitled "Natural and industrial arsenic" [5-7]; and "Environmental and industrial arsenic" [8,9]. Fowler [10] and Nriagu [11,12] have edited books devoted to the biological and environmental effects of arsenic. Chapters of interest appear in a book edited by Brinckman & Bellama [13], and some useful reviews have been published by Phillips [14]; Phillips & Depledge [15]; Maeda & Sakaguchi [16]; Cullen & Reimer [17]; Maeda [18]; Maeda and Takeshita [19]; and Takeshita and Maeda [20]. Arsenic compounds in marine environments are described in many of these books and reviews, but few describe those in freshwater environments.

3. Toxicity of Arsenicals

A number of different factors influence the toxicity of arsenicals, including the chemical forms in which arsenic is bound, physical form, mode of administration, species of organism, and criterion of toxicity. The chemical forms shown in Figure 1 differ in toxicity.

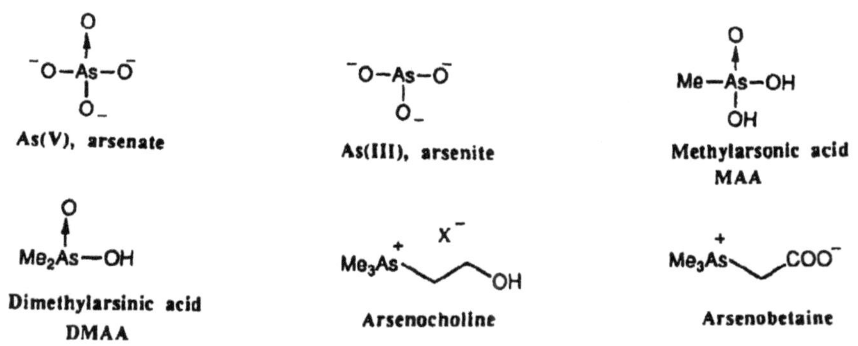

Figure 1. **Structures of some freshwater arsenic compounds**

3.1. INORGANIC ARSENIC COMPOUNDS

Trivalent arsenic compounds apparently exert their toxic effects chiefly by reacting with the sulfhydryl groups of vital cellular enzymes. Pyruvate dehydrogenase seems to be particularly vulnerable because it contains the thiol lipoic acid, which is especially reactive with trivalent arsenicals.

$$[\text{Enzyme}]\text{–SH} + \text{HOAs(OH)}_2 \longrightarrow [\text{Enzyme}]\text{–S–As(OH)}_2 + H_2O$$

The biochemical basis for the toxic action of pentavalent arsenic compounds is known with less certainty. A speculation is that such arsenicals compete with phosphate in phosphorylation to form unstable arsenate esters that spontaneously hydrolyze and thereby short-circuit energy-yielding bioenergetic processes. In other cases, the pentavalent arsenicals may exert toxic effects after being reduced in vivo to trivalent forms.

3.2. ORGANIC ARSENICALS

The most important members of this group are methylarsonic acid (MAA), dimethylarsinic acid (DMAA), and their salts. These compounds have been widely used as pesticides (primarily as herbicides and defoliants).

Both MAA and DMAA are well absorbed in animals and humans. By the oral route, these compounds produce symptoms of gastrointestinal irritation and renal and hepatic injury similar to some of the effects produced by inorganic arsenic, but the potency of these methyl derivatives is much lower.

Acute toxicity decreases in the following order: As(III) > As(V) > MAA > DMAA, suggesting that the acute toxicity of arsenic compounds diminishes with progressive methylation. Arsenobetaine [22], arsenocholine [23], trimethylarsine oxide [24], DMAA [24], MAA [24], and As(III) [22] have so far been fed to mice to estimate their toxicities, as summarized in TABLE 1. For arsenobetaine, the most ubiquitous arsenical in marine animals, an acute LD_{50} cannot be obtained because arsenobetaine administered orally to mice at a dose of 10 g per kg of body weight did not cause any toxic symptoms. The other trimethylated compounds, arsenocholine and trimethylarsine oxide, are 200 to 300 times less toxic than arsenic trioxide.

TABLE 1. LD_{50} (Oral administration to mice) of arsenic compounds

Arsenic compounds	LD_{50}(g/kg)	Reference
Arsenobetaine	>10.0	[22]
Arsenocholine	6.5	[23]
Trimethlarsine oxide	10.6	[24]
Dimethylarsinic acid (DMAA)	1.2	[24]
Methylarsonic acid (MAA)	1.8	[24]
Arsenic trioxide (As(III))	0.0345	[24]
Sodium arsenite (As(III) (rat)	0.0045	[24]

Buchet and coworkers administered oral doses (500 μg arsenic) of arsenite, MAA or DMAA to human volunteers [25]. The experimental results are shown in Figure 2. After four days, the excretion of urinary arsenic was 45, 78, and 75% of the dose of arsenite, MAA and DMAA, respectively. DMAA is excreted without any ansformation; no methylation or demethylation occurred. MAA also is not demethylated; but a small percentage (13%) can be transformed into DMAA. In the case of As(III) ingestion, the fraction of the dose excreted in four days amounts to only 45% but involves an important proportion (75%) of methylated metabolites.

4. Accumulation of Inorganic Arsenic Compounds by Freshwater Algae

Relatively little has been published on the concentrations of total arsenic in algae, especially in freshwater algae. However, algae should receive earnest attention because inorganic arsenic in the aqueous phase is first taken up by algae. This represents an

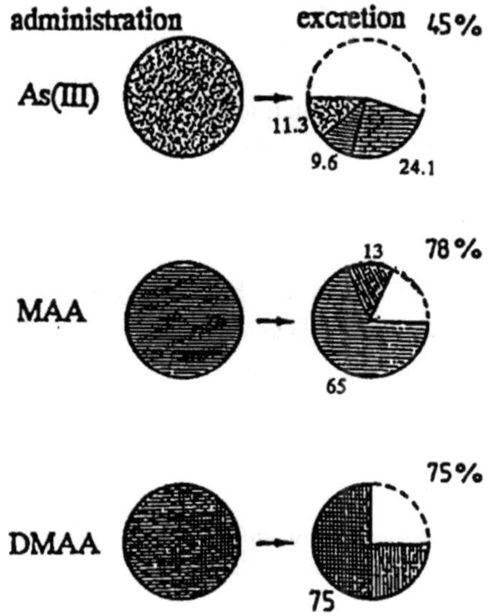

Figure 2. Urinary excretion of arsenic compounds within four days after oral administration of 500 mg arsenic as As(III) MAA or DMAA (Modified from Ref. [25])

important route for the introduction of arsenic into an aquatic food chain. Generally, the arsenic content in freshwater algae is lower than in marine algae. Some freshwater algae accumulate arsenic to a large degree. For example, Lunde [26] cultivated three freshwater algae (green alga: *Chlorella pyrenoidosa*; blue green alga: *Oscillatoria rubescence*; diatom: *Phaeodatylum tricornutum*) and three marine algae (green alga: *Chlorella ovalis*; diatoms: *Phaeodactylum* sp. and *Skeletonema costatum*) in freshwater and saltwater media containing radioactive arsenic ions. Enrichment of arsenic in the freshwater algae was 240 to 2800 times the concentration in the medium; in the marine algae, the enrichment was 710 to 2900 times. Thus the freshwater algae accumulated arsenic to a degree comparable with that of the marine algae tested.

Maeda and co-workers sampled algae collected from sites in Japan [27-29]. Most of the sites were abundant in As, Cu, Sn, or Au from old mines and in arsenic from a geothermal electric power plant. The algae were screened for arsenic tolerance. Some mixed algal systems grew well in a 50 to 2000 μg As(V)/g medium. The alga *Chlorella vulgaris* was isolated and found to survive in a medium containing 10,000 μg As(V)/g. Growth of *C. vulgaris* increased with increases in arsenate concentration up to 2000 As(V)/g, and accumulation of arsenic also increased with arsenic concentration in the medium. For example, in 100 μg As(V)/g medium the arsenic content in the cell was 20,000 μg/g; at stationary phase it was 10,000 μg/g dry weight. The maximum accumulation of arsenic from the 10,000 μg As(V)/g medium was about 50,000 μg/g dry cell weight at the phase of exponential growth.

Arsenite was more toxic to *C. vulgaris*, whose growth dropped off at concentration above the 10 μg As(III)/g medium and whose cells were cytolyzed at levels higher than 40 μg As(III)/g medium.

No As(V) was bioaccumulated by a cell which had been pretreated with dinitrophenol (respiratory inhibitor) or with heat. Little effect of sodium azide (photosynthesis inhibitor) on bioaccumulation was observed.

Phosphate competitively inhibited arsenic accumulation by *C. vulgaris* [28]. The cyanophyte *Synechococcus leopoliensis*, like a few other algae, was almost insensitive to arsenate and was unusual in that it showed high discrimination for phosphate; arsenate uptake was low and its fate was undetermined [30]. In other algae such as *Chlamydomonas* sp.[31] and *Phormidium* sp.[32], arsenate decreased phosphate uptake.

When *C. vulgaris* was precultured and acclimated in a medium containing As(V) at various levels, its arsenic uptake from the same media containing arsenic(V) at 1 and 100 $\mu g/cm^3$ was affected by the arsenic concentration in the preculture [28] (TABLE 2).

Table 2 shows that the higher the As(V) level in acclimation, the higher is the arsenic accumulation by *C. vulgaris*. Foster [33] reported there to be relationship between the tolerance and bioaccumulation of copper ion by *C. vulgaris* (the same species as the above). According to his results, the nontolerant strain was four times as sensitive to the copper ion but accumulated 5 to 10 times more metal than the tolerant strain, which had been acclimated to copper ion in a copper-polluted environment. These results imply that tolerance to the copper ion is attributable to copper exclusion by the cell. On the other hand, the arsenic-tolerant strain accumulated more arsenic than the nonacclimated strain. Therefore, different mechanisms are likely to be involved in the tolerances of *C. vulgaris* to arsenic and copper.

TABLE 2. Effects of arsenic acclimation on the arsenic bioaccumulation by *C. vulgaris* [28].

As(V) Level in Test Medium [mg As(V)/dm3]	Arsenic Accumulated in Cell at Stationary Phase (μg As/g dry wt)				
		As(V) Level in Acclimation Medium (mg As/dm3)			
	No acclimation	1	100	1000	5000
1	9	21	—	47	290
100	—	—	260	580	830

When *C. vulgaris* accumulating radioactive arsenic ($Na_2H^{74}AsO_4$) was transferred to an arsenic-free medium, the alga excreted 24 to 41% of its accumulated radioactive arsenic into the aqueous phase in six hours [34]. The alga excreted more arsenic into pure water than into a nutrient-enriched medium, and more in the dark than under illumination. These findings suggest that arsenic is readily excreted under conditions unfavorable for multiplication of *C. vulgaris.*

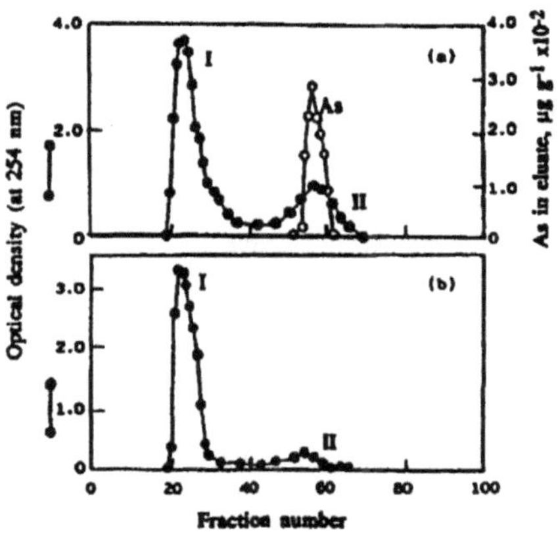

Figure 3. **Gel-filtration chromatography on Sephadex G-75 of proteins separated from arsenic-accumulated(a) and arsenic-free(b) *C.vulgaris* cells. Peaks I and II correspond to proteins having molecular weights of approximately 2 x 10⁴ and 3 x 10³, respectively. [35]**

Arsenic accumulated in living *C. vulgaris* cells was solvent-extracted with chloroform/ methanol (2:1), and the fractions were analyzed for arsenic [35]. A large part of the accumulated arsenic was localized in the extraction residues. The arsenic was found to be combined with a protein having a molecular weight of around 3000 in the arsenic-accumulated living cells (Figure 3). Analysis of the amino acids of the arsenic-bound protein showed that no metoallothionein-like protein was inductively biosynthesized in *C. vulgaris* on exposure to arsenic (TABLE 3). When the same alga was exposed to zinc and cadmium, metallothionein-like proteins were inductively synthesized in the alga [36,37]. It may not be necessary for the alga to biosynthesize metallothionein-like proteins, because the alga may have another detoxifying process for arsenic, such as methylation.

5. Biotransformation of Inorganic Arsenic Compounds by Freshwater Algae

Lunde [38] measured arsenic contents in lipids extracted from freshwater algae (*Chlorella ovalis*, *Phaeodactylum tricornutum*, and *Oscillatoria rubescens*), using the neutron activation technique. These algae bioaccumulated arsenic in the lipid phase at concentrations 0.5 to 5 μg As/g dry weight from enriched cultures containing 1 to 3 μg As/dm^3. These results suggest that the arsenic present in lipids from algae was organically bound, just as it is in oils extracted from more advanced marine organisms [26].

TABLE 3. Amino-acid composition of arsenicbound[35], arsenic-free[35], cadmium-bound[37] proteins, and data from literature.

	Amino–acid composition (mol%)			
Amino–acid	As–bound protein	As–free protein	Cd–bound protein	*Chlorella* in literature
Glycine	22.1	11.1	0.84	10.6
Alanine	14.1	19.8	5.42	8.76
Valine	6.64	7.12	2.40	6.42
Leucine	1.53	2.97	4.88	8.67
Isoleucine	1.36	2.00	4.06	5.61
Serine	11.4	5.04	6.79	5.13
Threonine	3.92	6.01	2.74	5.22
Cysteine	1.02	0.76	8.31	0.96
Methionine	0	2.70	0.43	1.30
Aspartic acid	0	6.08	0.03	9.14
Glutamic acid	14.1	8.78	7.37	9.36
Arginine	3.41	5.74	5.71	5.82
Lysine	3.07	3.59	1.34	6.38
Histidine	2.90	1.52	1.13	1.54
Phenylalanine	1.87	5.60	2.93	4.42
Tyrosine	0.51	2.35	3.37	2.50
Tryptosphan	0	0	1.54	1.08
Proline	12.1	8.71	42.1	7.05

When dried cells of *C. vulgaris* were digested with 2 M NaOH at 90 to 95 °C, the arsenic was released mainly as inorganic arsenic, which was identified as the hydride AsH_3 following $NaBH_4$ reduction [39]. Cells were also extracted with MeOH-$CHCl_3$ (2:1) to give (1) a residue, which contained 79% of the total arsenic, (2) a water-soluble fraction (4%), and (3) a lipid-soluble fraction (17%). Digestion (NaOH) of these three fractions, followed by $NaBH_4$ reduction, revealed that some methylation took place when the cells bioaccumulated arsenic. Most of the arsenic in the residue was present in inorganic forms, where it was bound in some way that prevented easy extraction by BAL (2,3-dimercapto-1-propanol). Similar results were obtained for other algae, namely, *Hydrocoleum* sp., *Phormidium* sp., *Microchaete* sp., and *Nostoc* sp., which bioaccumulated both arsenate and arsenite; the lipid fraction contained only some as yet uncharacterized methylarsenic compounds.

TABLE 4. Metabolism of methylated arsenic by C. *vulgaris* cultured for 7 days in MD medium containing 10 mg/dm^3 of methylated arsenic compounds[40]

Methylarsenic Species Added	Accumulated Arsenic Compounds[a] µg As/g dry wt (% of total arsenic)				
	Total	IA	MMA	DMA	TMA
Methylarsonic acid	50.9	0	24.0(47.2)	26.9(52.8)	tr[b]
Dimethylarsinic acid	60.0	0	0	60.0(100)	0
Arsenobetaine	58.1	0	0	tr	58.1(100)

[a] IA, inorganic arsenic; MMA, monomethylated arsenic; DMA, dimethylated arsenic; TMA, trimethylated arsenic.
[b] Detedcted but below detection limits

During the early exponential growth of *C. vulgaris*, a small part of arsenic accumulated in the cells was first transformed to monomethylarsenic compounds (MMA), and dimethylarsenic compounds (DMA); after a short time a part of these compounds was transformed to trimethylarsenic compound (TMA) [40]. When arsenic-free-living *C. vulgaris* was cultured in media containing methylated arsenic compounds (MAA, DMAA, and arsenobetaine), the methylated arsenics were similarly taken up by *C. vulgaris*, as shown in TABLE 4. About half of the MAA was transformed to DMA in the algal cells, but no demethylation occurred. When the cells were exposed to DMAA, only dimethylarsenic was accumulated. This means that neither methylation nor demethylation occurred during dimethylarsenic accumulation. In the case of arsenobetaine accumulation, no demethylation occurred. These experimental results revealed that *C. vulgaris* takes up not only inorganic arsenic but also methylated arsenic compounds and that methylarsenic compounds taken up may be further biomethylated but not demethylated. The dimethylarsenic compounds seem to be the most stable form of arsenic in the algal cells.

6. Biotransformation of Arsenic via the Experimental Freshwater Food Chain

Many papers on biotransformation of arsenic via marine food chains have been published. For example, three trophic levels of marine organisms (phytoplankton: *Dunaliella marina*; zooplankton: *Artemia salina*; and shrimp: *Lysmata seticaudata*) were tested for their arsenic metabolism [41]. The experimental results led to the conclusion that organic forms of arsenic in marine food webs were derived from *in vivo* synthesis by primary producers and were efficiently transferred along a marine food chain. The shrimp, the highest trophic level in this food chain, could not form organic arsenic by itself. In this case, arsenate taken up from water was converted largely to arsenite. Similar conclusions were obtained from experimental results on a phytoplankton-mussel

TABLE 5. Arsenic accumulation and metabolism in three-step freshwater food chain (phytoplankton *Nostoc* sp → shrimp *N. denticulata* →carp *C. carpio*) [51]

	Arsenic in organisms [μg As g^{-1} dry wt (%)]				
Organisms[a]	Total	IA	MMA	DMA	TMA
Nostoc sp.	74.3	736(99.1)	tr	6.7 (0.9)	—
N. denticulata	11.6	10.4(89.6)	—	0.6 (5.2)	0.6 (5.2)
C. carpio	5.2	4.1(78.8)	tr	0.4 (7.7)	0.7 (13.5)

Abbreviations: as shown in TABLE 4

[a] *Nostoc* sp. was cultured in MA medium containing 100 mg As(V) dm^{-3} for two weeks, *N. denticulata* was fed for seven days with the dried powder of the arsenic–dosed *Nostoc* sp., and *C. carpio* was fed for seven days with dried powder of the *N. denticulata* which had accumulated arsenic via arsenic–dosed *Nostoc* sp.

(*Mytilus galloprovincialis*)-crab (*Carcinus marnas*) system[42], a phytoplankton (*Fucus spiralis*)-grazer snail (*Littorina littoralis*)-carnivole snail (*Nucella laillus*) system [43,44], and a phytoplankton (*Dunaliella tertiolecta*)-lobster (*Homarus americanus* juveniles) system [45]. More information about biotransformation of arsenic in marine ecosystems is available in a review [46].

Very few papers on biotransformation of arsenic in freshwater ecosystems were available. Isensee et al.[47] examined the distribution of ^{14}C-labeled DMAA and dimethylarsine among freshwater organisms in a model ecosystem. Fish, *Daphnia magna*, snails, and algae were exposed to DMAA and dimethylarsine for 3, 29, 32, and 32 days. The freshwater organisms represented parts of two food chains: water —> algae —> snails and water —> diatoms, protozoa, and rotifers —> *Daphnia* —> fish. Lower food chain organisms (algae and *Daphnia*) bioaccumulated more DMAA and dimethylarsine, and the amounts accumulated indicate that DMAA and dimethylarsine do not show a high potential for biomagnification in the environment.

TABLE 6. Arsenic balance in *Moina* → *Poecilia* food chain[49].

			Arsenic in Organisms [ng (%)][a]			
Organism	Dry Mass (mg)	Total	Inorganic	MMA	DMA	TMA
175 *Moina* sp.[b]	0.875	60.2(100)	57.9(87.6)	tr[c]	8.2(12.4)	tr
Poecilia sp.	10	56(100)	9(15)	tr	1 (3)	46 (82)

[a]MMA, monomethylarsenic; DMA, dimethylarsenic, TMA, trimethylarsenic.
[b]Fed for seven days to *Poecilia* sp.

Recently Maeda and his co-workers reported experimental results on the transformation of arsenic via some experimental freshwater food chains [48-52]. In a three-step freshwater food chain (phytoplankton *Nostocm* sp. —> shrimp *N. denticulata* —> carp *C. carpio*), arsenic accumulation from food decreased one order of magnitude and the biomethylation ratio of the arsenic increased successively with elevation in the trophic level (see TABLE 5).

Similar results were obtained from an experiment in which another food chain (phytoplankton *C. vulgaris* and *Phormidium* sp. —> grazer *Moina* sp.—> carnivorous guppy *Poecilla* sp. juvenile) was examined[49]. TABLE 6 shows arsenic mass balances in the *Moina-Poecilla* food chain. Whereas the arsenic in the algae was almost all in inorganic arsenic (IA) form, 85% of the arsenic in the guppy was in DMA and TMA forms.

Figure 4 shows the relative concentration of arsenic species accumulated via the another food chain: alga *C. vulgaris* —> shrimp *N. denticulata* —> killifish *O. latipes* [52]. It shows that relative concentrations of DMA and TMA dramatically increased successively with elevation in the trophic level.

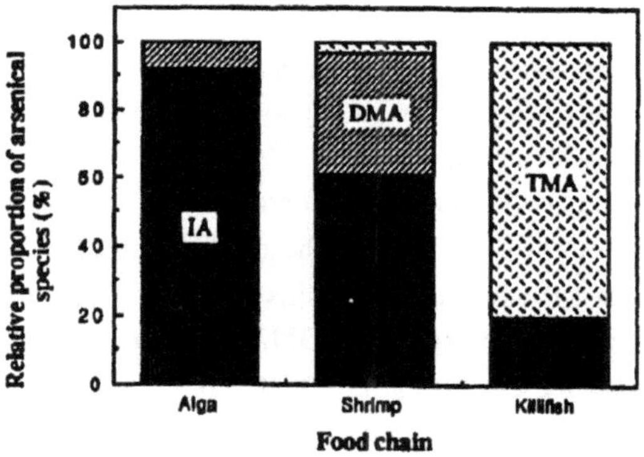

Figure 4. **Relative proportion of arsenic species in the three-step freshwater food chain (chlorella→shrimp→killifish) [52]. Chlorella was cultured in modified Detmer medium containing 100 μg As(V)/dm³ for one week, shrimp was fed for seven days with the dried powder of the arsenic-dosed Chlorella, and killifish was fed for seven days with the dried powder of the arsenic-dosed shrimp.**

■, IA; ■, MMA; ▨, DMA; ⊠, TMA.

Saito and Takahashi [53] analyzed arsenic in sweetfish living in a river polluted with arsenic from a hot spring (Takinoue Hot Spring, Iwate Prefecture, Japan). Arsenic concentrations were 14.3 μg As/g in sediment, 261 μg As/g in algae, and 2.24 μg As/g in river water. However, arsenic was not detected in the algae-eating sweetfish.
The results obtained from marine and from freshwater organisms closely resemble each other in that herbivorous and carnivorous organisms accumulated predominantly dimethylarsenic and trimethylarsenic, respectively. The experimental results from freshwater food chains showed that arsenic concentration in organisms decreases with elevation in the trophic level; the results from the marine food chain were the same [46].

7. Summary

Inorganic arsenic may undergo several biochemical transformations. Among them, the hydroxyl groups of arsenic acid, $AsO(OH)_3$, may be replaced by CH_3 groups to form $CH_3AsO(OH)_2$ (MAA), $(CH_3)_2AsOOH$ (DMAA). and $(CH_3)_3AsO$ (trimethylarsine oxide). These transformations usually occur biologically in aquatic environments. First, inorganic arsenic is incorporated by autotrophic organisms such as algae, and then it is transported through the food chain. During the transfer, the methylation of arsenic progresses in organisms. Part of the arsenic in freshwater organisms is combined with proteins, but arsenic-thionein is not biosynthesized [35,54]. Methylation of arsenic is probably the main detoxification process for organisms.

In general, organoarsenic species are dimethylarsenic compounds (DMA) in aqueous plants such as algae, and trimethylarsenic compounds (TMA) in mollusks, crustaceans, and fish, among marine organisms [14]. Similar results were obtained from freshwater organisms [49,50], but the detailed chemical species have not been confirmed.
Dimethylated arsenosugars are very widespread in marine macro-algae. Dimethylated arsenic is also found in freshwater algae, *Moina*, shrimp, and fish. The products from the chlorophyte *Rizoclonium* sp. are similar to those obtained from arsenoriboses [54]. But the chemical forms have not been confirmed yet.

Arsenobetaine, $(CH_3)_3As^+CH_2COO^-$, is a very widespread and possibly universal component of the arsenic in higher aquatic organisms in marine environments. Arsenobetaine is also found in higher organisms in brackish environments. Although trimethylated arsenic compounds have been found in higher organisms from freshwater environments [39,40,49,55], arsenobetaine has not yet been confirmed to exist in organisms in pure freshwater environments, except in shrimp.

In comparison of freshwater with marine organisms in their natural environments, there seems to be a difference in total arsenic concentration, the arsenic concentration in freshwater organisms being lower than in marine organisms [21]. However, organisms from both ecosystems resemble each other in that total arsenic concentration decreases and biomethylation of arsenic increases successively with elevation in the trophic level [46,49,50].

Although many heavy metals, such as mercury, cadmium, lead, and toxic organic compounds such as DDT are biomagnified via the food web, arsenic does not appear to be biomagnified in the freshwater environment as much as in the marine environment.

Arsenical compounds in freshwater algae are excreted in part and are soluble in water. The arsenic compounds may be transferred through surface waters by adsorption onto suspended sediments. The methylated arsenic compounds may be degraded finally into inorganic arsenic by bacteria under anaerobic conditions, as shown in marine bottom sediments [56,57]. The methylated arsenics may also be degraded photochemically; the effect of light is likely to be more significant in freshwater [58]. Consequently, arsenic becomes incorporated into the bottom sediments of lakes and ponds, where it remains as long as the overlying water remains aerobic.

8. References

1. National Research Council Committee on Medical and Biological Effects of Environmental Pollutants (1977). *Arsenic: Medical and biologic effects of environmental pollutants*. National Academy of Sciences, Washington, D.C., 332.
2. Anke, M., Schneider, H.-J., and Brückner, C., Eds. (1980). *Arsen.: 3rd. Spurenelement-Symposium* (1980). Friedrich-Schiller-University, Jena, 316 .
3. Lederer, W. H., and Fensterheim, R. J., Eds. (1983). *Arsenic: Industrial, biomedical, environmental perspectives*. Van Nostrand Reinhold Environmental Engineering Series, Van Nostrand Reinhold Company, New York, 443 .
4. Ishinishi, N., Okabe, S., and Kikuchi, T. Eds. (1985). Hiso: Kagaku, taisha, dokusei (*Arsenic: Chemistry, metabolism, and toxicity*). Koseisha-Koseikaku, Tokyo, 157 .
5. Irgolic, K. J., Kikuchi, T., Maeda, S., and Craig, P. J., Eds. (1988). Natural and industrial arsenic. *Appl. Organomet. Chem.*, Longman Scientific & Technical, Vol. 2, No. **4**, . 283-404.
6. Irgolic, K. J., Kikuchi, T., Maeda, S., and Craig, P. J., Eds. (1990). Natural and industrial arsenic. *Appl. Organomet. Chem.*, Wiley, Chichester. Vol. 4, No. **3**, . 181-295.
7. Maeda, S. and Craig, P. J., Eds. (1992). Natural and industrial arsenic. *Appl. Organomet. Chem.*, Wiley, Chichester. Vol. 6, No. **4**, 307-420.
8. Maeda, S. and Craig, P. J., Eds. (1994). Environmental and industrial arsenic. *Appl. Organomet. Chem.*, Wiley, Chichester. Vol. 8, No. **31**, . 165-283.
9. Maeda, S. and Craig, P. J., Eds. (1996). Environmental and industrial arsenic. *Appl. Organomet. Chem.*, Wiley, Chichester. Vol. 10, in press.
10. Fowler, B. A., Ed. (1983). Biological and environmental effects of arsenic: *Topics in environmental health*, Elsevier, Amsterdam, Vol. **6**, 281 .
11. Nriagu, J.o., Ed.(1994). *Arsenic in the environment*. Part I: Cycling and characterization, John Wiley & Sons, Inc., New York, 430 .
12. Nriagu, J.O.., Ed.(1994). *Arsenic in the environment.* Part II: Human health and ecosystem effects, John Wiley & Sons, Inc., New York, 293 .
13. Brinckman, F. E., and Bellama, J. M., Eds. (1978). *Organometals and organometalloids occurrence and fate in the environment.* ACS Symp. Ser., American Chemical Society, Washington, D. C., Vol. **82**, 447 .
14. Phillips, D. J. H. (1990). Arsenic in aquatic organisms: a review, emphasizing chemical speciation. *Aqua. Toxicol.*, **16**, 151-186.
15. Phillips, D. J. H., and Depledge, M. H. (1985). Metabolic pathways involving arsenic in marine organisms: A unifying hypothesis. *Mar. Environ. Res.*, **17**, 1-12.
16. Maeda, S., and Sakaguchi, T. (1990). Accumulation and detoxification of toxic metal elements by algae. In I. Akatsuka, Ed., *Introduction to applied phycology*. SPB Academic Publishing bv, The Hague, The Netherlands, 109-136.
17. Cullen, W. R., and Reimer, K. J. (1989). Arsenic speciation in the environment. *Chem. Rev.*, **89**, 713-764.
18. Maeda, S. (1983). Seibutsu-tainai no biryou-genso-kagakusyu to sono henkann (Chemical species of trace elements and the biotransformation in organisms). *Kagaku no Ryouiki* **37**, 606-610.
19. Maeda, S., and Takeshita, T. (1982). Shizen-kankyo ni okeru hiso-kagobutsu to seibutsu ni yoru hiso no taisya (Arsenic compounds in environment and metabolism of arsenic). *Kagaku-no-ryoiki*, **36**, 686-692.
20. Takeshita, T., and Maeda, S. (1980). Yuki-hiso-kagobutsu no taisya (Metabolism of organoarsenic compounds). *Yukagaku* (J. Jpn .Oil Chem. Soc.), **29**, 75-81.
21. Lunde, G. (1977). Occurrence and transformation of arsenic in the marine environment. *Environ. Health Perspect.* **19**, 47-52.

22. Kaise, T., Watanabe, S., and Itoh, K.(1985). The acute toxicity of arsenobetaine. *Chemosphere* **14**, 1327-1332.
23. Kaise, K., et al.(1992). Acute toxicity and metabolism of arsenocholine in mice. *Appl. Organomet. Chem.*, **6**, 369-373.
24. Kaise, K., Yamauchi, H., Horiguchi, Y., Tani, T., Watanabe, S., Hirayama. H., and Fukui, S. (1989). A comparative study on acute toxicity of methylarsonic acid, dimethylarsinic acid and trimethylarsine oxide in mice. *Appl. Organomet. Chem.*, **3**, 273-277.
25. Buchet, J.P. and Lauwerys, R. (1994). Interpretation of inorganic arsenic metabolism in humans in the light of observations made in vitro and in vivo in the rat. *Appl. Organomet. Chem.*, **8**, 191-196.
26. Lunde, G. (1973). The synthesis of fat and water soluble arseno organic compounds in marine and limnetic algae, *Acta Chem. Scand.*, **27**, 1586-1594.
27. Maeda, S., Kumamoto, T., Yonemoto, M., Nakajima, S., Takeshita, T., Higashi, S and Ueno, K. (1983). Bioaccumulation of arsenic by freshwater algae and the application to the removal of inorganic arsenic from an aqueous phase. Part I. Screening of freshwater algae having high resistance to inorganic arsenic. *Sep. Sci. Technol.*, **18**, 375-385.
28. Maeda, S., Nakashima, S., Takeshita, T., and Higashi, S. (1985). Bioaccumulation of arsenic by freshwater algae and the application to the removal of inorganic arsenic from an aqueous phase. Part II. By *Chlorella vulgaris* isolated from arsenic-polluted environment. *Sep. Sci. Technol.*, **20**, 153-161.
29. Maeda, S., Kumeda, K., Maeda, M., Higashi, S., and Takeshita, T. (1987). Bioaccumulation of arsenic by freshwater algae (*Nostoc* sp.) and the application to the removal of inorganic arsenic from an aqueous phase. *Appl. Organomet. Chem.*, **1**, 363-370.
30. Budd, K., and Craig, S. R. (1981). Resistance to arsenate toxicity in the blue-green alga *Synechococcus leopoliensis*. *Can. J. Bot.*, **59**, 1518-1521.
31. Planas, D. and Healey, F. P. (1978). Effects of arsenate on growth and phosphorus metabolism of phytoplankton. *J. Phycol.*, **14**, 337-341.
32. Maeda, S., Fujita, S., Ohki, A., Yoshifuku, I., Higashi, S., and Takeshita, T (1988). Arsenic accumulation by arsenic-tolerant freshwater blue-green alga (*Phormidium* sp.) *Appl. Organomet. Chem.*, **2**, 353-357.
33. Foster, P. L. (1977). Copper exclusion as a mechanism of heavy metal tolerance in a green alga. *Nature* (London), **269**, 322-323.
34. Maeda, S., Kusadome, K., Arima, H., Ohki, A., and Naka, K. (1992). Uptake and excretion of total inorganic arsenic by the freshwater alga *Chlorella vulgaris*. *Appl. Organomet. Chem.*, Vol. 6, No. **4**, 399-405.
35. Maeda, S., Arima, H., Ohki, A., and Naka, K. (1992). The association mode of arsenic accumulated in the freshwater alga *Chlorella vulgaris*. *Appl. Organomet. Chem.*, Vol. 6, No. **4**, 393-397.
36. Maeda, S., Mizoguchi, M., Ohki, A., and Takeshita, T. (1990). Bioaccumulation of zinc and cadmium in freshwater alga, *Chlorella vulgaris*. Part I. Toxicity and accumulation. *Chemosphere*, **21**, 953-963.
37. Maeda, S., Mizoguchi, M., Ohki, A., Inanaga, J. and Takeshita, T. (1990). Bioaccumulation of zinc and cadmium in freshwater alga, *Chlorella vulgaris*. Part II. Association mode of the metals and cell tissue. *Chemosphere*, **21**, 965-973.
38. Lunde, G. (1972). The analysis of arsenic in the lipid phase from marine and limnetic algae. *Acta Chem. Scand.*, **26**, 2642-2644.
39. Maeda, S., Wada, H., Kumeda, K., Onoue, M., Ohki, A., Higashi, S., and Takeshita, T. (1987). Methylation of inorganic arsenic by arsenic-tolerant freshwater algae. *Appl. Organomet. Chem.*, **1**, 465-472.
40. Maeda, S., Kusadome, K., Arima, H., Ohki, A. and Naka, K. (1992). Biomethylation of arsenic and its excretion by the alga *Chlorella vulgaris*. *Appl. Organomet. Chem.*, Vol. **6**, 407-413.
41. Wrench, J., Fowler, S. W., and Ünlü, M. Y. (1979). Arsenic metabolism in a marine food chain. *Mar. Pollut. Bull.*, **10**, 18-20.
42. Ünlü, M. Y. (1979). Chemical transformation and flux of different forms of arsenic in the crab *Carcinus maenas*. *Chemosphere*, **5**, 269-275.
43. Klumpp, D. W. (1980). Accumulation of arsenic from water and food by *Littorina littoralis* and *Nucella lapillus*. *Mar. Biol.*, **58**, 265-274.
44. Klumpp, D. W., and Peterson, p. J. (1981). Chemical characteristics of arsenic in a marine food chain. *Mar. Biol.*, **62**, 297-305.
45. Cooney, R. V., and Benson, A. A. (1980). Arsenic metabolism in *Homarus americanus*. *Chemosphere*, **9**, 335-341.
46. Andreae, M. O. (1983) Biotransformation of arsenic in the marine environment. In W. H. Lederer and R. J. Fensterheim, Eds., *Arsenic: Industrial, Biochemical, Environmental Perspectives*., Van Nostrand Reinhold, New York, . 378-392.
47. Isensee, A. R., Kearney, P. C., Woolson, E. A., Jones, G. E., and Williams, V. P. (1973). Distribution of alkyl arsenicals in model ecosystem. *Envir. Sci. Technol.*, **7**, 841-845.
48. Maeda, S., Inoue, R., Kozono, T., Tokuda, T., Ohki, A., and Takeshita, T. (1990a). Arsenic metabolism in a freshwater food chain. *Chemosphere*, **20**, 101-108.
49. Maeda, S., Ohki, A., Tokuda, T., and Ohmine, M. (1990). Transformation of arsenic compounds in a

freshwater food chain. *Appl. Organomet. Chem.*, **4**, 251-254.
50. Maeda, S., Ohki, A., Kusadome, K., Kuroiwa, T., Yoshifuku, I., and Naka, K. (1992e). Bioaccumulation of arsenic and its fate in a freshwater food chain. *Appl. Organomet. Chem..*, **6**, 213-219.
51. Maeda, S., Mawatari, K., Ohki, A., and Naka, K. (1993). Arsenic metabolism in a freshwater food chain: blue-green alga (*Nostoc* sp.) —> shrimp (*Neocaridina denticulata*) —> carp (*Cyprinus carpio*). *Appl. Organomet. Chem..*, **7**, 467-476.
52. Kuroiwa, T., Ohki, A., Naka, K. and Maeda, S.(1994). Biomethylation and biotransformation of arsenic in a freshwater food chain: Green alga (*Chlorella vulgaris*) —> shrimp (*Neocaridina denticulata*) —> killifish (*Oryzias latipes*). *Appl. Organomet. Chem..*, **8**,325-333.
53. Saito, K., and Takahashi, M. (1977). Heavy metals in sweetfish. *Iwate-ken Eisei Kenkyusyo Nempo*, **20**, 160-161.
54. Oladimeji, A. A. (1985). An arsenic-binding protein in rainbow trout. *Ecotoxicol. Environ. Saf.*, **9**, 1-5.
55. Maeda, S., Ohki, A., Miyahara, K., Higashi, S., and Naka, K.(1992). Biomethylation of methylated arsenic compounds by arsenic-resistant bacteria (*Klebsiella oxytoca* and *Xanthomonas* sp.). *Appl. Organomet. Chem.*, **6**, 415-420.
56. Hanaoka, K., Tagawa, S., and Kaise, T. (1992). The fate of organoarsenic compounds in marine ecosystems. *Appl. Organomet. Chem.*, **6**, 139-146.
57. Woolson, E. A., and Kearney, P. C. (1973). Persistence and reactions of ^{14}C-cacodylic acid in soils. *Environ. Sci. Technol.*, **7**, 47-50.
58. Brockbank, C. I., Batley, G. E., Low, G. K-C. (1988). Photochemical decomposition of arsenic species in natural waters. *Environ. Technol. Lett.*, 9, 1361-1366.

INVESTIGATIONS IN ORGANOARSENIC CHEMISTRY

IONEL HAIDUC AND LUMINITSA SILAGHI-DUMITRESCU
Facultatea de Chimie
Universitatea "Babes-Bolyai "
Cluj-Napoca, RO-3400 Romania

As a part of our continuing interest in organometallic chemistry [1], during the years our research group has been involved in the investigation of some organoarsenic compounds. This work was carried out mostly with sulfur-containing organoarsenic ligands (e.g. dithioarsinates) to serve for comparison with the dithiophosphinates, which were also a subject of intensive preoccupation in our laboratory (for reviews see [2,3]).

Our research in the area of organoarsenic compounds began in 1982, followed by the publication of a convenient method for preparing phenylarsenic(III) chlorides, from tetraphenyltin and arsenic trichloride [4]. At the time we did not realize that diphenylarsenic chloride is a warfare agent and we used the compound as starting material for diphenyldithioarsinic acid and other derivatives. The first interesting result followed, by discovering a sulfotropic molecular rearrangement of diphenylphosph(III)inyl diorganodithioars(V)inates, $Ph_2P\text{-}S\text{-}As(=S)R_2$, with formation of diorganoarsenic(III) diphenyldithiophosph(V)inates, $Ph_2P(=S)\text{-}S\text{-}AsR_2$ [5]. This was followed by investigation of the sulfotropic rearrangement of tetra-organoarsinedisulfides, $R_2As\text{-}S\text{-}S\text{-}AsR_2 \rightarrow R_2As(=S)\text{-}S\text{-}AsR_2$ [6]. A large number of organosilicon-, -germanium, -tin, -and -lead derivatives of dimethyl- and diphenyl-dithioarsinic acids were prepared [7,8], as well as several inorganic main group metal dithioarsinates [9]. For one of these, dimethyltindimethyldithioarsinate, $Me_2Sn(S_2AsMe_2)_2$, the crystal and molecular structure was determined [8].

We paid some particular attention to organoantimony(III) dithioarsinates. Thus, diphenylantimony(III) diphenyldithioarsinate, $Ph_2Sb\text{-}S\text{-}As(S)Ph_2$, was synthesized, and unexpectedly, the crystal structure determination revealed that this is a ladder-type tricyclic dimer, containing an eight-membered $Sb_2S_4As_2$ ring contour [10]. By contrast, the dimethyldithioarsinate analogue, $Ph_2Sb\text{-}S\text{-}As(S)Me_2$, was found with the aid of single crystal X-ray diffraction to be a coordination polymer associated via bridging dimethyldithioarsinate ligands [11]. Such associations were recently described as supramolecular [3]. Another example of supramolecular self-assembly

J.F. Bunnett and M. Mikołajczyk (eds.), Arsenic and Old Mustard: Chemical Problems in the Destruction of Old Arsenical and 'Mustard' Munitions, 149–150.

was found in an unusual tetranuclear adduct $[Ph_2As(=O)(OH)\cdot Ph_2As(=S)(OH)]_2]$, associated through hydrogen bonds [12].

A collaboration with Dr. R. Cea-Olivares (Mexico) led to the synthesis and structural investigation of phenoxarsin-10-yl dithioarsinates, which provided the first example of a dimeric organoarsenic(III) 1,1-dithiolate associated through As ··· S secondary bonding [13]. In the 1-oxa-4,6-dithia-5-arsocane and 1,3,6-trithia-2-arsocanedithiocarbamates, it was found that competition between endocyclic (transannular) and exocyclic secondary As ··· S bonds occurs, which influences the overall structure [14].

In relation with these investigations some theoretical studies were also performed; thus, we reported a normal coordinate analysis of the vibrational spectrum of the dimethyldithioarsinato anion, $Me_2AsS_2^-$[15].

We believe our results contribute to the conclusion that organoarsenic chemistry can be an attractive area of chemical and structural investigation, in spite of the known toxicity of such materials. We have found, in collaboration with toxicologists (unpublished results), that organoarsenic-sulfur derivatives are less toxic than their halogen or oxygen analogues. We have had no safety or toxicity problems during work with these compound in our laboratory. This is probably due to the fact that arsenic is very thiophylic and sulfur-containing ligands are not displaced by biological sulfur-containing molecules.

References

1. I. Haiduc and J. J. Zuckerman (1985) *Basic Organometallic Chemistry*, Walter de Gruyter, Berlin, New York.
2. I. Haiduc and D. B. Sowerby (1996), *Polyhedron*, **15**, 2469.
3. I. Haiduc (1997) *Coord. Chem. Rev.*, **158**, 325.
4. L. Silaghi-Dumitrescu and I. Haiduc (1983), *Syn. React. Inorg. Metal-org. Chem.* **13**, 475.
5. L. Silaghi-Dumitrescu and I. Haiduc (1983) *J. Organomet. Chem.* **252**, 295.
6. L. Silaghi-Dumitrescu, I. Silaghi-Dumitrescu and I. Haiduc (1989) *Rev. Roumaine Chim.* **34**, 305.
7. (a) I. Haiduc and L. Silaghi-Dumitrescu (1982) *J. Organomet. Chem.*, **225**, 225;
   (b) L. Silaghi-Dumitrescu and I. Haiduc (1983) *J. Organomet. Chem.* **259**, 65.
8. L. Silaghi-Dumitrescu, I. Haiduc, JWeiss (1984) *J. Organomet. Chem.* **263**, 159.
9. L. Silaghi-Dumitrescu (1986) L. Avila-Diaz and I. Haiduc, *Rev. Roumaine Chim.* **31**, 335.
10. C. Silvestru, L. Silaghi-Dumitrescu, I. Haiduc, M. J. Begley and D. B. Sowerby (1986) *J. Chem. Soc. Dalton Trans.*, 1031.
11. D. B. Sowerby, M. J. Begley, L. Silaghi-Dumitrescu, I. Silaghi-Dumitrescu and I. Haiduc (1994) *J. Organomet. Chem.* **469** ,45.
12. L. Silaghi-Dumitrescu, M. N. Gibbons, I. Silaghi-Dumitrescu, J. Zukerman-Schpector and D. B. Sowerby (1996) *J. Organomet. Chem.* **517**, 101.
13. R. Cea-Olivares, J. G. Alvarado, G. Espinosa-Perez, C. Silvestru and I. Haiduc (1994) *J. Chem. Soc. Dalton Trans.*, 2191.
14. R. Cea-Olivares, M. R. Estrada, G. Espinosa-Perez, I. Haiduc, P. Garcia y Gracia, M. Lopez-Cardoso, M. Lopez-Vaca and A. M. Cotero-Villegas (1995) *Main Group Chem.* **1**, 159.
15. I. Silaghi-Dumitrescu, L. Silaghi-Dumitrescu and I. Haiduc (1982) *Rev. Roumaine Chim.* **27**, 911.

HPLC-ICP-MS METHODS FOR THE DETERMINATION OF INORGANIC AND ORGANIC ARSENIC COMPOUND

W. GOESSLER, D. KUEHNELT AND K.J. IRGOLIC
Institute for Analytical Chemistry, Karl-Franzens-Universität Graz, Austria

Concentrations of total arsenic in soil and water samples contaminated with old Arsenical Munitions are not very useful to characterize the potential risks. Knowledge of which arsenic compounds are present in such samples is absolutely necessary to define toxicity. The identification of arsenic compounds requires a separation step combined with a detection step. For separation, gas chromatography and high performance liquid chromatography are widely used. Atomic absorption spectrometers, inductively coupled plasma optical emission spectrometers, and inductively coupled plasma mass spectrometers may serve as arsenic-specific detectors.

Because the concentrations of arsenic compounds in soil and water samples are frequently low, only techniques of high sensitivity that enable accurate quantitative measurement to low concentration limits can be used for the determination of arsenic compounds. The hydride generation technique has low detection limits, but can be used only for the detection and quantification of arsenic compounds that are easily volatile or can be reduced to easily volatile arsines. High performance liquid chromatography will separate all arsenic compounds that can be detected by ICP-MS in the column effluents. Arsenic compounds can still be quantified, when at least 0.1 ng of arsenic is present in the injected volume (~ 100 mL).

Anion-, cation-exchange, and ion-pair chromatographic methods for the determination of arsenic compounds in biological samples were developed. A VG PlasmaQuad Turbo Plus inductively coupled plasma mass spectrometer served as arsenic-specific detector. To improve the detection limits, a hydraulic high-pressurenebulizer was substituted for the normal pneumatic nebulizer. Chromatographic conditions for the separation of inorganic and organic arsenic compounds are identified in Table 1. Characteristic chromatograms are presented in Figs. 1, 2 and 3.

J.F. Bunnett and M. Mikołajczyk (eds.), Arsenic and Old Mustard: Chemical Problems in the Destruction of Old Arsenical and 'Mustard' Munitions, 151–154.

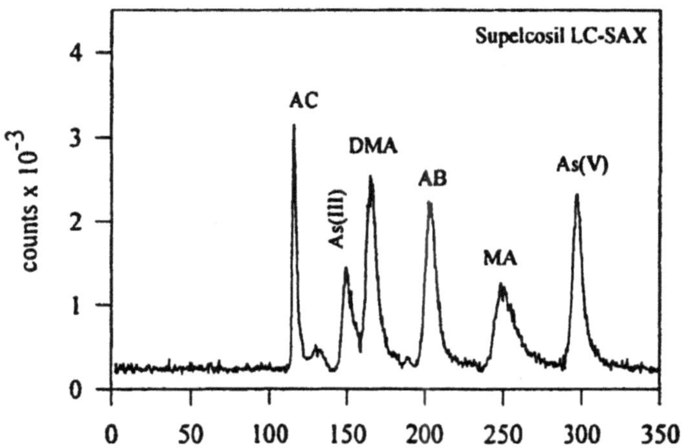

Figure 1. Chromatogram obtained with a solution of arsenate, arsenite, methylarsonic acid, dimethylarsinic acid, arsenobetaine, and arsenocholine (0.5 ng of As each species) in distilled water on a Supelcosil LC-SAX anion-exchange column (mobile phase 30 mM $NH_4H_2PO_4$ with 1 % methanol at pH 3.75, injection volume 0.100 cm^3, flow rate 1.5 cm^3 min^{-1})

Table 1

Column	Mobile Phase	pH	Compounds separated
Hamilton PRP-X100	30 mM $NH_4H_2PO_4$	6.0	As(III), DMA, MA, As(V)
Supelcosil LC-SAX	30 mM $NH_4H_2PO_4$	3.75	AC, As(III), AB. DMA. MA. As(V)
Supelcosil LC-SCX	20 mL Pyridine	3.0	AB, TMAO, AC, Tetra
PRP-1	10 mM 2-hydroxy-5-sulfobenzoic acid	3.5	TMAO, Tetra, AC

Abbreviations: As(III): arsenite; As(V): arsenate; DMA: dimethylarsinic acid; MA: methylarsinic acid; AC: arsenocholine; AB: arsenobetaine; TMAO: trimethylarsine oxide; Tetra: tetramethylarsonium ion

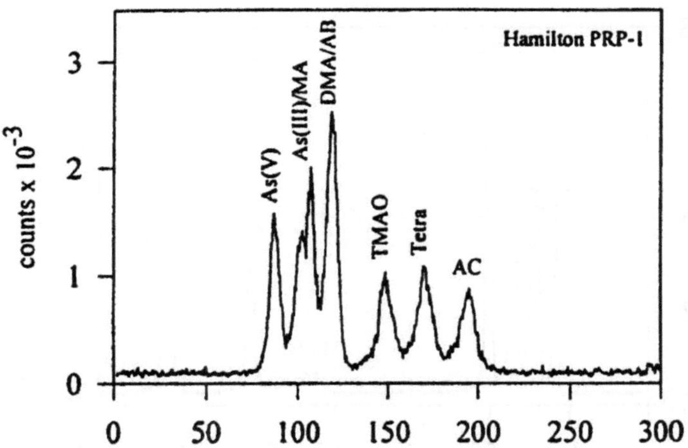

Figure 2. Chromatogram obtained with a solution of arsenate, arsenite, methylarsonic acid, dimethylarsinic acid, arsenobetaine, trimethylarsine oxide, arsenocholine, and tetramethylarsonium iodide (0.5 ng of As each species) in distilled water on a Hamilton PRP-1 reversed-phase column (mobile phase 10 mM 2-hydroxy-5-sulfobenzoic acid) at pH 3.5, injection volume 0.100 cm^3, flow rate 1.5 cm^3 min^{-1}.

The combination of the hydraulic high-pressure nebulizer with the inductively coupled plasma mass spectrometer enables the separation and detection of arsenic at the sub-ng/mL level. These methods can certainly be used to identify and quantify arsenic compounds in soil samples suspected of contamination with arsenic-containing chemical warfare agents.

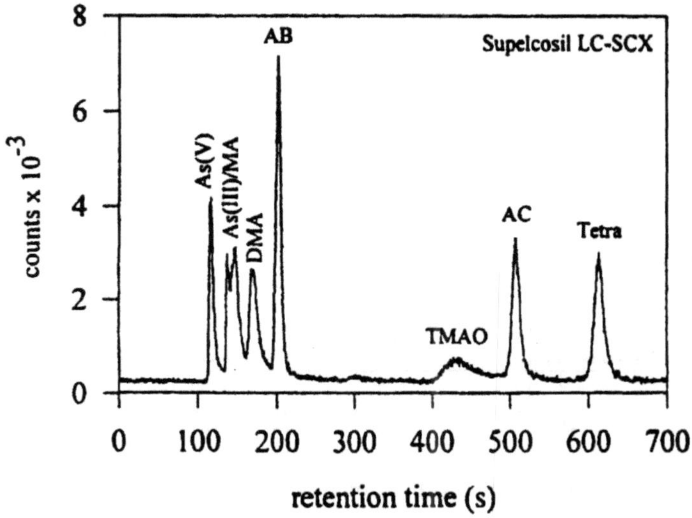

Figure 3. Chromatogram obtained with a solution of arsenate, arsenite, methylarsonic acid, dimethylarsinic acid, arsenobetaine, trimethylarsine oxide, arsenocholine, and tetramethylarsonium iodide (0.5 ng of As each species) in distilled water on a Supelcosil LC-SCX cation-exchange column (mobile phase 20 mM pyridine at pH 3.0, injection volume 0.100 cm^3 , flow rate 1.5 cm^3 min^{-1}).

A METHOD FOR MUTUAL DISPOSAL OF OLD CHEMICAL WEAPONS

ALEXANDER L.CHIMISHKYAN
D. Mendeleev University of Chemical Technology
Moscow, Russia

It is known that the hydrogen chloride effects destruction of Adamsite with formation of diphenylamine and arsenic trichloride.

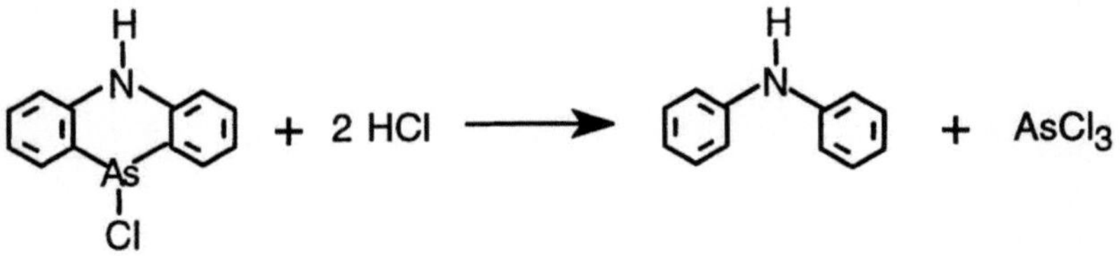

Also known is that the reaction of diphenylamine with another toxic agent, phosgene, is a method for production of acridone-9, a compound from a series of naphthoacridinic dyes. [4]

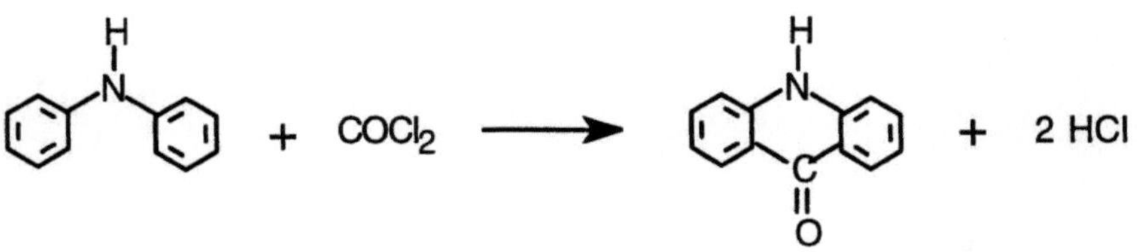

We have used these two reactions for the mutual disposal of Adamsite and phosgene. Our preliminary investigation has shown that the best results can be obtained when phosgene is passed through an Adamsite/paraffin melt. At the high temperature of the process (190-200°C), arsenic trichloride vapours are removed from the reaction mass, where acridone-9 accumulates. We have also found that tetraphenylurea is present in the reaction mass. The process guarantees a high yield of arsenic trichloride.

J.F. Bunnett and M. Mikołajczyk (eds.), Arsenic and Old Mustard: Chemical Problems in the Destruction of Old Arsenical and 'Mustard' Munitions, 155–156.

A mixture of phosgene and hydrogen chloride can be formed by hydrolysis of tetrachloromethane in the presence of ferric chloride. [5]

$$CCl_4 + H_2O \xrightarrow{FeCl_3} COCl_2 + 2\,HCl$$

We are now investigating a comprehensive one-pot method for transformation of Adamsite into acridone-9 in which phosgene and HCl are produced in the reaction mass by controlled tetrachloromethane hydrolysis.

1. References

1. K. Lohs (1958), *Synthetische Gifte*, Ministerium fuer Nationale Verteidigung, Berlin,
2. S. Franke (1967), *Lehrbuch der Militaerchemie, Bd.I*, Deutscher Militaer-verlag, Berlin.
3. V. Gormay, V. Shanovalov, A. Shantrokha et al. (1994),*Russian Chem. J.* **2**, 39.
4. C.Nenitescu (1963) ,*Chimie organica*, Vol II, Editura Tehnica, Bucuresti.
5. M.Hill (1960), *J. Org. Chem.* **25**, (7), 1115.

DESTRUCTION OF ADAMSITE BY SODIUM

M. SOKOLOWSKI* AND E. BILGER#
* *Military Institute of Chemistry and Radiometry, Wieczorkiewicza 100, PL - 00910 Warsaw, Poland*
# *Dr. Bilger Umweltconsulting GmbH, Brunnenstrasse 11 - 13, D - 63579 Freigericht-Somborn*

The Sodium-Technology is applicable to the chemical destruction of chemical weapons (1,2) such as Adamsite. The aim of the commercially approved Sodium-Technology is to react chemicals in a liquid, dissolved or gaseous state with a special form of metallic high surface sodium, either on solid supports („black sodium") or as a dispersion in oil (particle sizes of 5-10 μm) to yield inorganic salts of sodium and small organic molecules of little or no toxicity.

Adamsite can be destroyed by the reaction with sodium at low and even ambient temperatures within a short time. Most of the originally organic bonded arsenic is transformed into inorganic (ionic) arsenic and therefore can be precipitated as an insoluble sulfide. In addition an insoluble residue assumed to be Phenarsazin, according to recent MS data, is also found.

Sodium attacks directly the different bonding systems in organoarsenic molecules dissolved in tetrahydrofuran or xylene. As a first step, the formation of Na_3As is likely under such inert conditions.

After completion of Adamsite decomposition, water containing sodium sulfide (dilute) is added in the closed compartment to initiate the hydrolysis of sodium organic compounds and the precipitation of As_2S_3, which is formed via $As(OH)_3$ in the reaction:

$$Na_3As + 6\,H_2O \longrightarrow 3\,NaOH + As(OH)_3 + 3H_2$$

Then $$2\,As(OH)_3 + 3\,Na_2S \longrightarrow As_2S_3 + 6\,NaOH$$

J.F. Bunnett and M. Mikołajczyk (eds.), Arsenic and Old Mustard: Chemical Problems in the Destruction of Old Arsenical and 'Mustard' Munitions, 157–158.

Hydrogen gas can then be released for optional further treatment. Since As^{+3} partly will oxidise to As^{+5} in aqueous solution, a mixture of insoluble sulfides As_2S_5 and As_2S_3 will be isolated.

As the final step, the aqueous solution is subjected to UV-activated oxidation. Any residual organic structure will decompose upon UV-activated oxidation with hydrogen peroxide. As final products in the water phase, only ions of Na^+, Cl^-, and NO_3^- with traces of $SO_4^=$ are expected. The nature of the As-containing residual compounds is under further analytical investigation.

Experience in the use of the Sodium-Technology for destruction of highly toxic chlorinated dioxins and furans indicates that any existing Adamsite can be treated in a mobile detoxification unit within a short time.

References

1. Patent DE 4022526
2. Patent DE 19501597

ELECTROCHEMICAL AND BIOLOGICAL APPROACH TO THE DESTRUCTION OF LEWISITE AND 'MUSTARD'

ALEXANDER M. BORONIN[a], VALENTIN G. SAKHAROVSKI[a], IVAN I. STAROVOITOV[a*], KONSTANTIN I. KASHPAROV[a], VALERY N. SHVETSOV[b], KSENIJA M. MOROZOVA[b], IGOR A. NECHAEV[b], VLADIMIR I. TUGOSHOV[c], NIKOLAI P. KUZMIN[a] AND ALEXANDER I. KOCHERGIN[c]

a. Institute of Biochemistry and Physiology of Microorganisms, Russian Academy of Sciences, 142292 Pushchino, Moscow Region, Russia;
b. Research and Engineering Institute on Water Supply, Sewage Systems, Hydraulic Engineering Units and Engineering Hydrogeology, 119826 Moscow, Komsomolsky prospect, 42, Russia;
c. State Institute of Technology of Organic Synthesis, 412680 Volsk-17, Saratov Region, Russia

Ecologically benign destruction of Lewisite and 'Mustard' has been achieved by electrochemical pretreatment of their neutralization products followed by biological oxidation of the waste stream.

In line with the International Chemical Weapons (CW) Convention (1993), the development of environmentally benign technology for CW destruction has become a pressing problem. The CW stockpiles in Russia contain both nerve (Sarin, Soman and VX) and blister ('Mustard' and Lewisite) agents. Until recently, the Russian CW disposal program envisaged a two stage process of chemical neutralization of CW agents and then incineration of waste streams. However, organized opposition to this CW disposal program has arisen because this technology is not environmentally benign.

Alternative to incineration for destroying CW agents is biologically based technology; biotechnological waste stream treatment may accomplish complete mineralization of chemicals. We examined the possibility of using a biological process as the final stage of waste stream treatment from degradation of Lewisite and 'Mustard'.

J.F. Bunnett and M. Mikołajczyk (eds.), Arsenic and Old Mustard: Chemical Problems in the Destruction of Old Arsenical and 'Mustard' Munitions, 159–162.

Lewisite is hydrolyzed in water according to the equation:

$$Cl\text{-}CH{=}CH\text{-}AsCl_2 + H_2O \rightarrow \underset{\mathbf{1}}{Cl\text{-}CH{=}CH\text{-}As{=}O} + 2HCl$$

Direct biodegradation of **1** is not promising due to the high content of As in a reaction solution.

To remove arsenic from a reaction solution we used electrolysis for destruction of **1** followed by electrocoagulation (EC) of arsenic compounds.

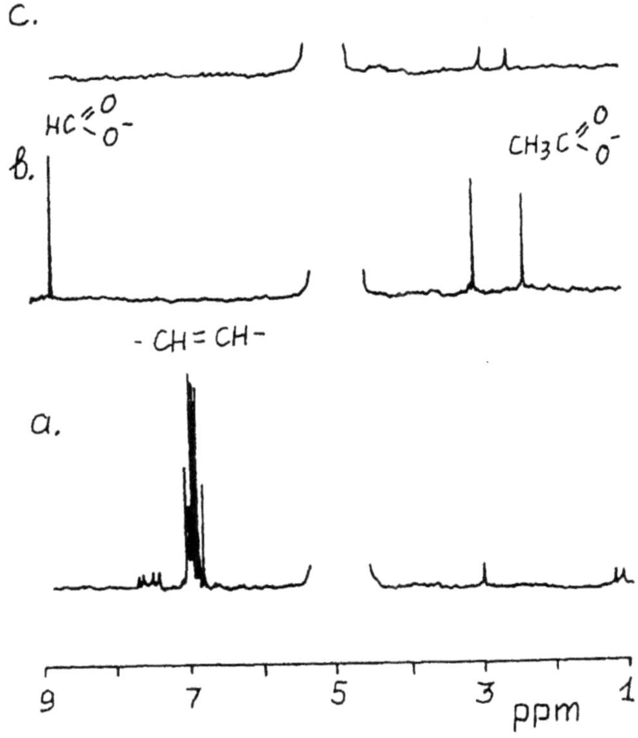

Figure 1. 400 Mhz[1]H NMR spectra of (a) 2–chlorovinyl arsine oxide; (b) products of its electrochemical destruction after bioutilization. Solvent - D_2O.

The products of electrochemical destruction of **1** in a nondiaphragm electrolyzer were formate, acetate (Fig.1), arsenous and arsenic acids, and arsenic. Electrolysis of **1** (2 g/l) was performed for 2 h at anode current density 0.01 A/cm^2 and the progress of the reaction was followed by monitoring the disappearance of the shoulder at 2440 nm in the UV spectrum of the solution. The reaction solution after electrolysis was subjected to EC with aluminum electrodes. The method of nondiaphragm electrolysis allows the decrease of arsenic content three-fold (from 1.02 up to 0.32 g/l). Arsenic concentration was further decreased by three orders of magnitude to a range of 0.41 mg/l by EC.

The products of electrochemical pretreatment of **1** were oxidized in a biosorber by microorganisms immobilized on activated carbon. Arsenic salts after EC may be used as a secondary raw material.

Thus, a combined approach for Lewisite destruction was proposed, which includes: 1) water hydrolysis of Lewisite to **1**; 2) electrochemical degradation of **1** to low-molecular organic acids; 3) separation of arsenic salts by EC; and 4) biodegradation of the organic products of electrolysis.

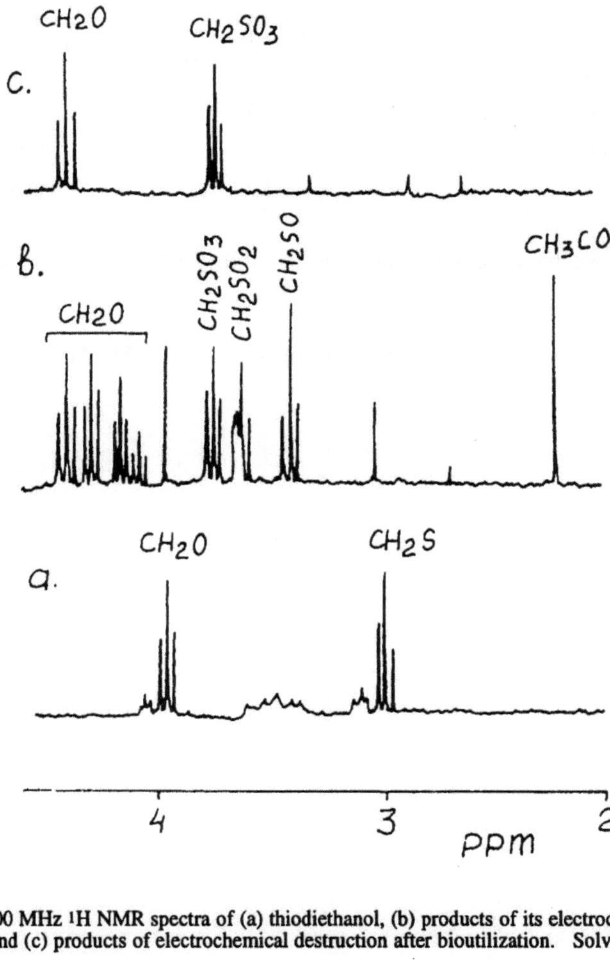

Figure 2. **400 MHz ^{1}H NMR spectra of (a) thiodiethanol, (b) products of its electrochemical destruction, and (c) products of electrochemical destruction after bioutilization. Solvent: D_2O**

For 'Mustard' (HD) detoxification, alkaline hydrolysis was used. The main reaction product was 2,2'-thiodiethanol (2) (Fig. 2, zone a). We used electrochemical pretreatment of HD detoxification products to prepare them for biological processing. The main products of electrochemical oxidation of **2** were 2-hydroxyethanesulfonate (**3**), bis(2-hydroxyethyl) sulfoxide (**4**), bis(2-hydroxyethyl) sulfone (**5**) and acetic acid (**6**) (Fig. 2, zone b).

The products of electrochemical oxidation of **2** were oxidized by activated sludge in biosorber (Table 1; Fig. 2, zone c).

TABLE 1. The products of electrolysis of 2, 2'-thiodiethanol (g/1)

	Bioutilization of the Products	
	aerobic process	anaerobic process
2-hydroxyethanesulfonate (1.39 g/l)	81.5	87.0
bis(2-hydroxyethyl) sulfoxide (0.58 g/l)	63.5	78.3
bis(2-hydroxyethyl) sulfone (0.06 g/l)	77.3	91.7
acetate (0.1 g/l)	>99.0	>99.0

In conclusion, we have demonstrated the feasibility of ecologically benign HD destruction.

This work was supported partly by NATO Linkage Grant DISRM. LG 951060.

REDUCTIVE DECOMPOSITION OF DEPOSITS IN OLD 'MUSTARD' MUNITIONS

JOSEPH F. BUNNETT
University of California
Santa Cruz, CA 95064; USA

The nature of the gelled or even solid deposits in decades-old 'mustard' tanks or munitions is uncertain. Some possibilities, which may occur alone or in combination:

(a) polysulfonium salts, from alkylation of one molecule by another, etc.

(b) poly(ethylenesulfide)s, $—CH_2–CH_2–S–CH_2–CH_2–S–CH_2–CH_2–S—$, resulting from chloride ion attack on polysulfonium salts

(c) elemental sulfur, an impurity in Levinstein 'mustard'

(d) di- or oligosulfide polymers

(e) iron compounds, incorporating iron from container corrosion

All these structural types are vulnerable to attack by strong reducing agents. Sodium in liquid ammonia cleaves dialkyl sulfides and dialkyl disulfides: [1,2]

$2\ Na + R–S–R + NH_3 \longrightarrow R–S^-\ Na^+ + RH + NaNH_2$

$2\ Na + R–S–S–R \longrightarrow 2\ RS^-\ Na^+$

Electrochemical studies [3,4] imply that sulfonium salts are similarly cleaved:

$R_3S^+\ X^- + 2\ Na + NH_3 \longrightarrow R_2S + RH + NaX + NaNH_2$

Sulfur reacts with sodium: $S_n + 2n\ Na \longrightarrow n\ Na_2S$

Iron salts are reduced by sodium: $FeS + 2\ Na \longrightarrow Fe + Na_2S$

Alkyl halides are cleaved by sodium: $R–X + 2\ Na + NH_3 \longrightarrow RH + NaX + NaNH_2$

In the laboratory, conducting such reactions in liquid ammonia is attractive because solutions of sodium in ammonia are stable for hours or days. Although liquid ammonia boils at -33°C, it can readily be kept in the liquid state either in a Dewar flask or in common glass flasks fitted with well-type condensers cooled by solid CO_2. In industry, chemical syntheses are sometimes conducted in liquid ammonia.

Factory-type installations utilizing sodium in liquid ammonia to clean out old 'mustard' munitions with gelled or solid deposits should therefore be feasible. After a munition had been opened and drained of the 'mustard' that would pour out, liquid ammonia could be added followed by pieces of sodium metal to react with and destroy the deposits. A munition so treated would be safe to ship to a furnace to burn out traces remaining.

In a field situation, sodium/ammonia treatment would be inconvenient. A promising alternative is use of sodium in ethanol (grain alcohol). Sodium reacts with ethanol to release hydrogen gas:

$$2\ CH_3CH_2OH + 2\ Na \longrightarrow 2\ CH_3CH_2ONa + H_2$$

J.F. Bunnett and M. Mikołajczyk (eds.), Arsenic and Old Mustard: Chemical Problems in the Destruction of Old Arsenical and 'Mustard' Munitions, 163–164.

It has however long been known sodium in ethanol effects "dissolving metal reductions", such as the classical Bouveault-Blanc reduction of carboxylic esters to alcohols, in good yields:

$$R\text{-}\overset{\overset{\displaystyle O}{||}}{C}\text{-}OCH_3 + 4\,Na + 4\,CH_3CH_2OH \longrightarrow R\text{-}CH_2\text{-}OH + CH_3OH + 4\,CH_3CH_2O^- Na^+$$

The mentioned reactions with sodium in liquid ammonia are expected also to occur with sodium in ethanol. Research is however needed to confirm this expectation.

Whether in ammonia or ethanol, the active reducing agent is the solvated electron, released from sodium: $Na \longrightarrow Na^+ + e_s^-$. In ammonia solvent, e_s^- has a lifetime of hours or days, but in ethanol its half life is only about six microseconds.[5] Its reactivity with many organic compounds is however so high that it largely reacts with them before decomposing. For example,with propyl chloride (1 M) e_s^- has half life 1.4 nanoseconds.[1] Actually, rate constants for solvated electron reactions in water [6] are more available; the following half-lives (with 1 M solutes) are for reactions with e_s^- in water: dimethyl sulfide, 35 ns; dimethyl sulfoxide, 0.4 ms; cystine (a disulfide), 0.3 ns; iron(III) cation, 0.014 ns; and methyl propanoate, 7.7 ns. Most of these half lives are orders of magnitude shorter than the half life of the solvated electron in ethanol. Reactions should be complete while solvated electron are still present.

Because sodium metal reacts violently with liquid water and even with water vapor of the air, it would need to be carefully packed for use in the field. Adequate packing should be practical. One package might contain enough for use on a single munition, or several kilograms for use on several in quick succession.[7] CAUTION: although sodium in ethanol may react mainly with the partially solidified (as well as liquid) 'mustard', some reaction with ethanol to produce inflammable hydrogen gas will occur.

Laboratory research is needed to confirm that compounds modeling the various suggested structures for gelled or solidified 'mustard' are indeed reactive with sodium in ethanol as expected. If those expectations are fulfilled, development work on sodium reactions with CW agents, in ammonia and in ethanol, should follow.

References.

1. Williams, F. E.; Gebauer-Fuelnegg, E. (1931) *J. Am. Chem. Soc.*, 53, 352.
2. Watt, G. W. (1950), *Chem. Rev.*, 46, 317.
3. Colichman, E. L.; Love, D. L. (1953) *J. Org. Chem.*, 18, 40.
4. Lüttringhaus, A. and Machatzke, H. (1964) *Justus Liebigs Ann Chem.*, 671, 165. The sulfonium salts studied were closely related to 'mustard'.
5. Freeman, G. R. (1974) *Radiation Chemistry of Ethanol,* NSRDS-NBS 48, National Bureau of Standards, Washington, DC.
6. Anbar, M., Bambenek, M. and Ross, A. B. (1973) *Selected Specific Rates of Reactions of Transients in Water in Aqueous Solution.* 1. Hydrataed Electron, NSRDS-NBS 43, National Bureau of Standards, Washington, DC.
7. Complete reaction of sodium with 'mustard' (monomer), to form NaCl, Na_2S and ethane, requires 116 grams of sodium per 100 grams of 'mustard'.

USE OF 'MUSTARD' AMINOLYSIS PRODUCTS AS CATALYSTS FOR POLYURETHANE FOAM PRODUCTION

A. L. CHIMISHKYAN
S. I. ORLOV
T. S. SEREBRYAKOVA
Russian D. Mendeleev University of Chemical Technology
Moscow A-47, Russia 123047

'Mustard' is the chemical warfare agent for which methods of destruction are the most extensively developed.

High temperature, direct incineration of 'mustard' is characterized by high efficiency. Under optimum conditions, the extent of destruction is as high as 99.9999%. A perceived risk that dioxins and benzofurans may be emitted is however a drawback of this technology. [1]

High temperature pyrolytic methods, including pyrolysis in a plasma, guarantee the absence of dioxins among decomposition products. [2] Successful realization of these technologies depends on the availability of equipment of appropriate design.

Numerous known methods destroy 'mustard' by interaction with detoxicants. Among these are a variety of hydrolysis procedures. [3] For efficient destruction hydrolysis by water alone is conducted at temperatures from near the boiling point of water to several hundred degrees (in the case of supercritical water). [4,5] Lower temperatures suffice when alkalis [6] and catalysts [7] are present.

The main factor that limits use of 'mustard' hydrolysis technologies is its very low solubility in water (0.0043 mol/l at 25 °C).

Destruction can also be effected by materials, such as hydrogen peroxide and calcium hypochlorite ("chloride of lime"), that have long been used for decontamination purposes. [3, 6] Because of their highly corrosive properties, utilization of these detoxicants is disfavored on account of waste disposition considerations.

Methods for detoxification of 'mustard' by conversion to useful, non-toxic polymers have been developed. This approach is attractive as a way to offset the huge expense of other technologies. Polycondensation of 'mustard' with alkali metal polysulfides results in polymers resembling polysulfide rubbers ("thiokols").[6] Inasmuch as the quantity of 'mustard' subject to destruction in Russia is about 700 tons while the annual production of polysulfide rubbers amounts to thousands of tons, this methodology would not make a major contribution.

J.F. Bunnett and M. Mikołajczyk (eds.), Arsenic and Old Mustard: Chemical Problems in the Destruction of Old Arsenical and 'Mustard' Munitions, 165–168.

Reactions of 'mustard' with amines potentially can form products of special value to industry even if not needed in large quantities. Because weapons-grade 'mustard' is a complex mixture of chloroalkyl sulfides [6], the materials formed in reactions with amines may be mixtures. For some purposes, they may nevertheless be quite suitable.

In particular, the products formed by reaction of 'mustard' with amines have promise to serve as catalysts for reactions to form polyurethane (PU) foams. Such catalysts are not required in large quantities. Thus the limited supply of "raw material" to form them should satisfy the needs of industry for many years. Obtaining useful materials from a notorious remainder of war preparedness will not only reduce the cost of destruction of 'mustard', but will be spiritually rewarding.

PU foams are prepared by reactions of diols with toluene 2,4-diisocyanate; the reactions are catalyzed by bases, including tertiary amines. We have investigated reactions of 'mustard' with primary and secondary amines. We have also explored the catalytic activity of the resulting amines for PU production.

The use of ethanolamine for detoxification of CW agents, including organophosphonates [9], has received considerable study [1,3,6]. N-(2-hydroxyethyl)thiomorpholine (I) is formed by its reaction with 'mustard' (Scheme I). This product has been shown by toxicological studies to be of class III danger (Russian official category; $LD_{50} \sim 7000$ mg/kg) [1].

It is known that reactions of 'mustard' with ammonia and primary amines affords thiazanes, heterocycles in which the amine nitrogen is a ring member [8]. We examined the reaction of 'mustard' with dimethylamine, seeking to obtain open-chain amino sulfides. When the amine was in excess, bis(2-dimethylaminoethyl) sulfide (II) was formed [see Scheme I). Its oxygen analog, bis(2-dimethylaminoethyl) ether, is widely used in the manufacture of PU foams [10].

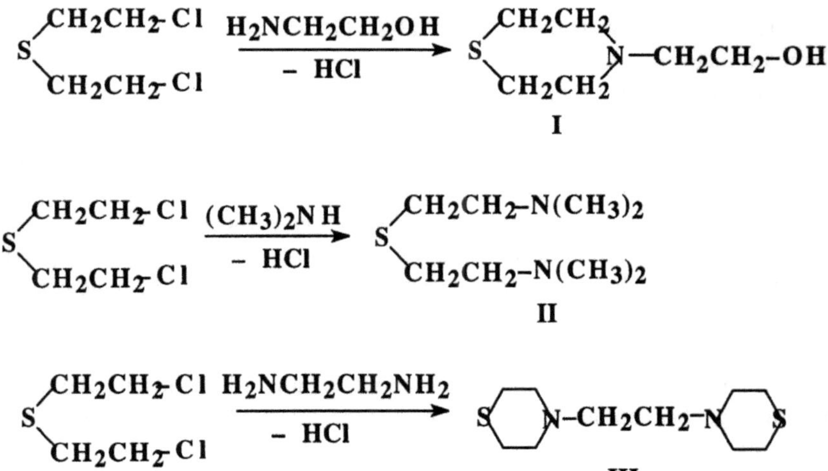

Scheme I

N,N,N',N'-Tetramethylethylenediamine is an effective catalyst for PU foam form-ation but is not used because it is too volatile. Reaction of 'mustard' with ethylene-diamine promised to form a bis-tertiary amine of similar chemical character but lower volatility. The expected reaction (Scheme I) to form 1,2-ethane–N,N'–bis(thiomopholine) (III) occurred in modest yield.

1. Experimental

'Mustard' for this work was prepared by a Clark method [11]. The properties of the amines used dictated the conditions for each reaction. Thus, in the case of dimethylamine, the reaction was carried out in water at 140-150 °C under pressure of 20-25 atm., in an autoclave. Ethanolamine for reaction with 'mustard' contained 10% of ethylene glycol, to moderate the exothermic effect [1]. The reaction was conducted at about 105-110 °C. In the case of ethylenediamine, no solvent or diluent was needed; the reaction proceeded well at 90-110 °C. When cyclic compounds I and III were prepared, the order of addition of reactants was important. Addition of amines to 'mustard' favored formation of the products mentioned above. The formation of undesired, bioactive quaternary ammonium compounds was thereby avoided. All the reactions were exothermic; temperature was controlled by the rate of addition of amine.

Changes in reaction conditions allow various products to be formed from primary amines and 'mustard'. Excess amine tends to favor bis-amines resulting from replacement of two chlorines by two amine nucleophiles. The generation of biologically active quaternary ammonium salts is promoted by having 'mustard' in excess, and by solvents of high polarity. The formation of thiazane derivatives is favored by having the amine and 'mustard' present in equimolar concentrations, and by higher temperatures.

In all cases, the hydrochlorides of the indicated amine products were present at reaction end; they were converted to the free bases by treatment of the reaction mixtures with aqueous NaOH. Products I, II and III were purified by extraction with chlorofom followed by distillation. bis(2–Dimethylaminoethyl) sulfide (II), b.p. 93–95°C (10 torr), was obtained in 75% yield. The yield of N–(2–hydroxyethyl)thiomorpholine (I) was 60%, after double distillation; it had b.p. 123-129 °C (2 torr) and m.p. 35-37 °C. These properties coincide with those previously published for I, prepared by reaction of thiomorpholine with 2-chloro-ethanol [12]. 1,2-Ethane-N,N'-bis(thiomorpholine) (III) crystallized from its concentrated chloroform extract, and was recrystallized from benzene; m.p. 103-105 °C; yield 15 - 20%. III has also been prepared by reaction of thiomorpholine with 1,2-dibromoethane [13].

The catalytic activity of products I, II and III, as well as of some comparison amines, was assessed by viscometric measurements. A polyether triol, Laprol 5003-2 B-10[1] was combined with commercial toluene diisocyanate T 80 (Bayer AG)[2] plus some water and acetone, for simulation of foam production conditions. The molar ratio of hydroxyl to isocyanate groups was 3/1. The amines tested for catalytic activity were also present; they were 1,4-dimethylpiperazine (DMP), bis(2-morpholino-ethyl) ether (BMDEE) and 1,4-diazabicyclo[2.2.2]octane (DABCO). The concentrations of amines were in the range 0.006 to 0.4 M. (DABCO is used in industrial production of PU foams, while DMP and BMDEE are used as co-catalysts.)

2. References

1. Lugansky, I.N., Sheluchenko, V.V., Krotovich, I.N., *et al.* (1994) Main technological and economic aspects of mustard destruction, *Russian Chemical J.* **38,** No 2, 34-36 (rus.).
2. Baker, S., Derre, P, and Schtelt E. (1993). Safe destruction of the high toxic substances, *Russian Chemical J.* **37,** No 3, 29-33 (rus.).
3. Umerov, I.A., Kuznetsov, B.A., Krotovich, I.N., *et al.* (1993) Methods of destruction and utilization of lewisite and mustard stored, *Russian Chemical J.* **37,** No 3, 25-29 (rus.).
4. *Report of the NATO advanced research workshop: Destruction of military toxic wast, (1994)* Naaldwigk, Netherland, 22-27 May 1994.
5. Koshelev, V.M., Zhdanov, V.A., Shuvalov, A.A., *et al.* (1995) American development of methods of a chemical weapons destruction, *Russian Chemical J.* **39,** No 4, 31-36 (rus.).
6. Soborovsky, L.Z. and Epstain, G.Yu. (1938) *Chemistry and technology of military chemical agents,* State Defensive Industry Publisher, Moscow-Leningrad (rus.).
7. Sherbakova, L.F., Shantroha, A.V., Gormay, V.V., *el al.* (1993) in: *Ecological problems of a chemical weapons destruction,* Theses of the reports. A scientific-technological conference, 10-11 June 1993, Volsk, pp. 13-14 (rus.).
8. Franke, S. (1967) *Lehrbuch der Militärchemie. Band I,* Deutscher Militärverlag, Berlin.
9. Vasil'ev, I.A., Shvyryaev, B.V., Liberman, B.M., *et aL* (1995) Kinetics and mechanism of interaction Sarin with ethanolamine and mathematical simulation of detoxication reactor, *Russian Chemical J.* **39,** No 4, 5-9 (rus.).
10. Saunders, J.H., Frisch, K.C. (1962) *Polyurethanes, Chemistry and Technology, Part I, Chemistry,* Interscience, New York.
11. Alexandrov, V.N. and Emelyanov, V.1. (1990) *Military chemical agents,* Military Publisher, Moscow (rus.).
12. Beilstein. *Handbuch der Organischen Chemie,* EIII/EIV, C 27, p. 645.
13. Asinger, F., Saus, A., Hartig, J., *et al.* (1979) Zur Kenntnis der Reactionsfähigkeit des unsubstisuren Thiomorpholins und alkylsubstituieren Thiomorpholine, *Monatsh. Chem.* **10,** No 4, 767-789

---

1 Laprol 5003-2 B-10 is a block copolymer, average molecular weight 5000. Its content of ethyleneoxy units, contributed by ethylene oxide, is about 10%.

2 Toluene diisocyanate T 80 is stated by Bayer AG to comprise an 80/20 blend of its 2,4- and 2,6-isomers, respectively.

BREAKDOWN OF SULFUR 'MUSTARD' BY PHASE TRANSFER CATALYSED HCl ELIMINATION

A Potential Destruction Method For 'Mustard' Stocks

ERNST - CHRISTIAN KOCH[1,2]
ABC- und Selbstschutzschule
Mühlenweg 12, 87527 Sonthofen

1. Introduction

The Chemical Weapons Convention [1] requires the signatory countries to destroy all chemical weapons in their property, stockpiled as well as buried old chemical ammunition. No directions have been set up on how these weapons have to be destroyed, except that mere dumping and open air incineration are prohibited. Thus great efforts were made to develop suitable technologies for large scale destruction of CWs [2-5].

It is essential that the destruction of chemical agents not lead to substances which can easily be reconverted into CWs [6]. Common CWs include nerve agents like sarin (GB), soman (GD) and VX, and blistering agents like 'mustard' (HD) or Lewisite (L). While nerve agents (G and V agents) easily hydrolyse, 'mustard' (HD) is only slightly soluble in water (~1g/l at ambient temperature) [7] and hence hydrolytic decomposition of 'mustard' is impeded. It is true that HD can be hydrolyzed in water containing up to 3.8 % 'mustard' at temperatures around 84-86°C, but this procedure only works in case of distilled 'mustard' [8]. Thickened 'mustard' agents and tarry material in HD [9] do not dissolve in hot water. Work up of these agents is thus complicated.

Therefore 'mustard' is often incinerated. The drawbacks of incineration are low public acceptance of any form of incineration and, far more important, insufficiency of incineration capacity. Hence many chemical [10-13], physical [14-17] and even biological [18] methods to destroy 'mustard' are now being tested.

---

[1] Present address: Chemical R&D Department, Piepenbrock Pyrotechnik Gmbh, 67307 Göllheim, Germany.

[2] Dedicated to Professor Cornelius G. Kreiter on the occasion of his 60th birthday.

J.F. Bunnett and M. Mikołajczyk (eds.), Arsenic and Old Mustard: Chemical Problems in the Destruction of Old Arsenical and 'Mustard' Munitions, 169–174.

2. Destruction of 'Mustard'

Methods for 'mustard' destruction have been investigated for a long time [19, 20]. From the standpoint of chemical reactivity, one may consider decomposition by nucleophilic attack at a chlorine-bearing carbon atom as well as oxidation of the sulfur atom. Since 'mustard' sulfoxide and sulfone are still potent vesicants, oxidation will be disregarded. Mustard decomposition via nucleophilic attack is as old as the use of 'mustard' as a vesicant.

At the end of World War I the hydrolysis of 'mustard' was intensively studied. It was found that hydrolysis was influenced by a multitude of parameters, temperature, duration of contact and alkalinity being the most important ones [21-25]. Other investigations concerned the reaction of 'mustard' with alkali metal sulfides to give nontoxic 1,4-dithiane [19,20,50]. During World War II hydrolysis was reinvestigated in many countries. It was then found that the reaction proceeds stepwise via a unimolecular mechanism [25-26] involving a cyclic sulfonium ion intermediate [27]. In recent years mustard hydrolysis has been thoroughly studied by Appler, Reichert, Tilley and Yang [28-35, 51,52].

3. Problem

For reasons of military tactics, 'mustard' was always used in mixtures with thickeners (chlorinated rubber {'Tornesit' $(C_{10}H_{12}Cl_8)_n$}, wax {'S-Wachs'}, polyacrylates {'UNCON 75-M-50 000'}). The resulting viscous 'mustard' was known in German as Zählost, ZL. The mixtures sometimes contained also other CW agents such as Lewisite or HN [46] or even anthracene oil or CA to camouflage its distinct smell [49]. These types of 'mustard' are only slightly soluble or even insoluble in water. Other constituents were decomposition products, byproducts such as polysulfides, and tarry material, all of which decrease solubility of the agent in aqueous media. Since the problem of low solubility has been recognized [2,4], attempts have been made to hydrolyze 'mustard' in solvents other than water [48].

It is possible to dissolve 'mustard' and Zählost in organic solvents like benzene or toluene [47]. Hydrolysis in these media is however decreased by the reduced polarity of the solvent. In a less polar solvent the rate for the ion pair return step is significantly increased and thus hydrolysis is impeded. Also the repression of 'mustard' hydrolysis by chloride ion must be considered [10].

A solution to this problem could be phase transfer catalysis. As it is known from the investigations by Martinetz et al. [36-40], nucleophilic substitution of the chlorine in 'mustard' by cyanide proceeds smoothly under phase transfer conditions. Also acetate/halogenide exchange could be observed [36-40]. One should therefore expect hydrolysis of 'mustard' under phase transfer conditions to proceed in a similar way.

4. Results and Discussion

Treatment of a solution of distilled 'mustard' (**1**) in toluene at 60°C with aqueous sodium hydroxide (**2**) in the presence of catalytic amounts of tetrabutylammonium bromide (**3**) leads after 2 hours to complete decomposition of bis(2-chloroethyl) sulfide (**1**) and formation of divinyl sulfide (**4**) [41,42] and 2-chloroethyl vinyl sulfide (**5**) [43] (in ratio **4**/**5** = 3:1) as the only reaction products. After 3h, **5** had been completely converted into **4**. There has been no indication for intermediate formation of 2-chloroethyl 2'-hydroxyethyl sulfide (**6**) nor for 2,2'-thiodiethanol ("thiodiglycol") (**7**).

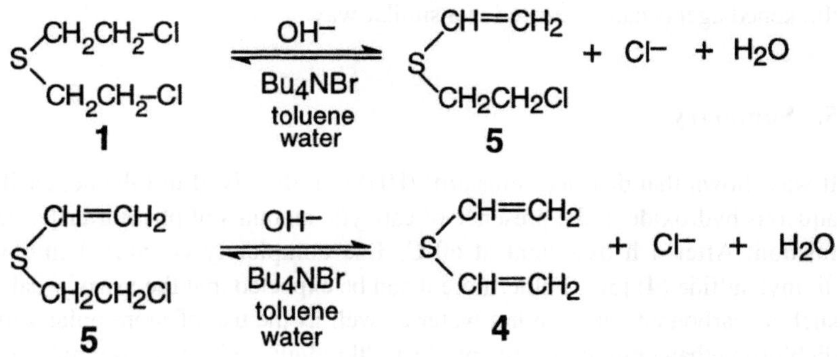

Scheme 1

The conditions (low polar solvent, strong nucleophile) should favour the E1 mechanism. Identification of **5** indicates the stepwise character of the reaction (Scheme 1).

The conversion of sulfur 'mustard' (**1**) to divinyl sulfide (**4**) has been known since the 1920's [44] and has been recently reported again to occur on treatment of **1** with superoxide in the presence of cobalt(II) tetrasulfonated tetraamino phthalocyanine complexes [53].

As the hydroxide initiated decomposition of 'mustard' proceeds in organic solvents of low polarity such as toluene, it should be possible to treat thickened 'mustard' in a similar way. This could be a good starting point for large scale destruction of 'mustard' munitions. A hypothetical treatment of divinyl sulfide is shown in Scheme 2.

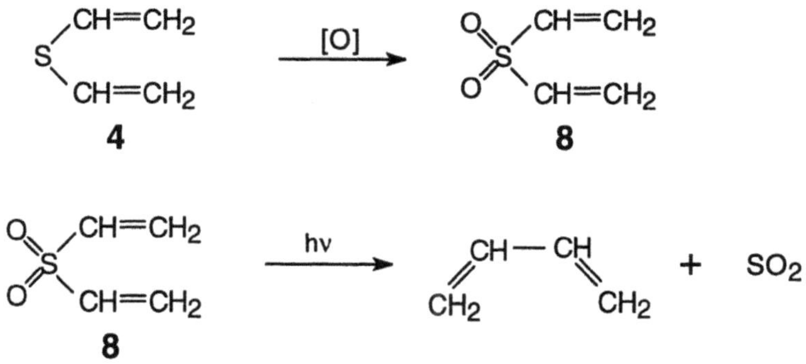

Scheme 2

The known oxidation of **4** should yield the sulfone (**8**) [54-55]; this in turn could be converted by irradiation to SO_2 and butadiene as the only reaction products. Further experiments are in progress, to reveal the reaction mechanism and to test whether thickened agents can be treated in a similar way.

5. Summary

It was shown that distilled 'mustard' (HD) (**1**), dissolved in toluene, easily reacts with aqueous hydroxide in the presence of catalytic amounts of phase transfer catalysts in E1 fashion. After 4 h treatment at 60°C, **1** is completely converted into weakly toxic divinyl sulfide (**4**) [56]. Futhermore it can be expected that the use of weak nucleophiles such as carbonates or even hot water as well as the use of more polar solvents such as dichloromethane might give the products "thiodiglycol" (7) and divinyl sulfone (8).

6. Acknowledgement

Valuable discussions held with Oberstleutnant W. Bretschneider as well as preparative assistance by H. Wassmer and T. Brucker are gratefully acknowledged. The author thanks ABC- und Selbstschutzschule for provision of sulfur 'mustard' (HD).

7. Instrumentation

IR spectra were obtained from toluene solutions in cells with NaCl windows and a 0.2 mm path length in the region 3500-500 cm^{-1} with an FT-IR spectrometer model 1760 X (*Perkin Elmer*). GC-MS measurements were made on a MM1 (*Bruker Franzen*) system using a DB-1 capillary column, 20 m long. Liquid samples were adsorbed on Tenax GK 26-02 (Günther Karl GmbH).

8. Experimental

A two necked 100 ml reaction vessel with reflux condensor, magnetic stirring bar and thermometer was charged with 25 ml toluene and 0.001 mol (0.32 g) tetrabutylammonium bromide (**3**) (Fluka >98%). The solution was stirred at ambient temperature until all crystals had dissolved. Then 0.02 mol (3.2 g, 2.5 ml) of distilled bis-(2-chloroethyl) sulfide (**1**) was added and the solution was stirred again. Finally 25 ml of 8 N aqueous sodium hydroxide (0.2 mol) and 8 g of NaOH (Merck, 99%) in 25 ml water were placed in the vessel. The mixture was stirred (600 rpm) and heated to 60°C using a heating jacket.

Divinyl sulfide (**4**): MS: m/z: 86(69%), 85(100%), 71(15%), 60(30%), 59(82%), 58(86%), 57(31%), 45(75%), 34(10%); IR: 1588 (vs), 957(s), 896(sh), 870(m), 589(w) cm^{-1}.

2-Chloroethyl vinyl sulfide (**5**): MS: m/z: 122(30%), 85(18%), 73(93%), 63(19%), 62(20%), 61(18%), 60(100%), 59(77%), 58(59%), 57(28%), 48(19%), 47(29%), 46(23%), 45(95%); IR: 629 (m) cm^{-1}.

9. References

1. Übereinkommen über das Verbot der Entwicklung, Herstellung, Lagerung und des Einsatzes chemischer Waffen und über die Vernichtung solcher Waffen. 13. Januar 1993, in *Internationales Recht der Rüstungsbeschränkung*, **13**. Lfg., 1994 Berlin.
2. Bunnett, J. F. (1995) *Pure & Appl. Chem.* **67**, 841.
3. Report of the NATO Advanced Research Workshop on the Destruction of Military Toxic Waste, Naaldwijk, The Netherlands, 22-27 May 1994.
4. Report of the NATO ARW on Chemical Problems Associated with Old 'Mustard' and Arsenical Munitions, Lodz, Poland, 17-19 March 1996. This volume.
5. Report of the 3rd Workshop of the HSP Pugwash Study Group on the Implementation of the Chemical and Biological Weapons Convention, Naaldwijk, The Netherlands, 19-21 May, 1995.
6. ECCN 4798B (the Australia-Group list for precursor chemicals of chemical weapons).
7. Stöhr, R. (1985), *Chemische Kampfstoffe und Schutz vor chemischen Kampfstoffen*, p. 150, Militärverlag der DDR, Berlin.
8. Blades, T. this volume.
9. Tarry material is probably a 'mustard' polymerization product formed via catalysis by ferric chloride and hydrochloric acid, which are byproducts in technical mustard.
10. Yang, Y.C., Baker, J.A., Ward,J.R. (1992), *Chem. Rev.*, **92**, 1729.
11. Umyarov, I.A., Kuznetsov, B.A., Krotovich, I.N., et al. (1993) *Ross. Khim. Zh.*, **37**, 25.
12. Boronin, A.M., Sakharovskii, V.G., Starovoitov, I. I., Kashparov, K.I., Zyakun, A.M., Shevtsov,V.N., Morozova, K.M., Nechaev, I. I., Tugushov, VI., Kuzmin, N.P., Kochergin, A.I. (1996) *Prikl. Biokhim. Mikrobiol.* **32**, 56.
13. Koch, M., this volume.
14. Yang,Y.C., Buchanan, J.H., Rohrbaugh, D.K., Szafraniec, L.L., Beaudry, W.T. May (1989) The Photocatalytic Oxidation of HD on Titania Surfaces; CRDEC-SP-005.,
15. J .H. Buchanan, Rohrbaugh, D.K., Szafraniec, L.L., Beaudry, W.T., Yang,Y.C., Photocatalytic Air Oxidation of HD and VX. In 1989 Inhouse Laboratory Independant research (ILIR).
16. Fox, M. A., Kim, Y. S., Abdel-Wahab, A. A., Dulay, M. (1990), *Catal. Lett.* **5**, 369.
17. Fox, M.A. (1983), *Acc. Chem. Res.* **16**, 314.
18. Harvey, S.P., Height, J.J., Szafraniec, L.L., Beaudry, W.T., DeFrank, J. J., Valdes, J.J., Chakrabarty, A.M. (May 1991) Biodegradation of Mustard, in *Abstracts of 91st Annu. Met. Am. Soc. Microbiol.*, Dallas, TEX, Abstr. Q 119.

19. Prentiss, A.M. (1937) *Chemicals in War - A Treatise on Chemical Warfare*, p. 1984, McGraw Hill Book, New York, 1937.
20. Hanslian, R., (1937) *Der Chemische Krieg, Teil 1 Militärischer Teil*, Verlag von E. S. Mittler&Sohn, Berlin.
21. Hopkins, E.F. (1919) *J. Pharmacol.* **12**, 393.
22. Hopkins, E.F., *ibid.* p. 403.
23. Rona, P. (1921) *Z. f. ges. Exp. Med.* **13**, 16.
24. Wilson, R.E., Fuller, E.W., Schur, O.M. (1922) *J. Am. Chem. Soc.* **44**, 2867.
25. Peters, R.A., Walker, E. (1923) *Biochem. J.* **17**, 260.
26. Mohler, H., Hartnagel, J., Sorge, J. (1940) *Mitt. Lebensm. Hyg.* **31**, 115; Mohler, H., Hartnagel, J. (1941) *Helv. Chim. Acta* **24**, 564.
27. Bartlett, P. D., Swain, C. G. (1949) *J. Am. Chem. Soc.* **71**, 1406.
28. Yang, Y.C., Szafraniec, L.L., Beaudry, W.T., Ward, J.R. (1988) *J. Org. Chem.* **53**, 3293.
29. Yang, Y.C., Szafraniec, L.L., Beaudry, W.T., Ward, J.R. (1987) *J. Org. Chem.* **52**, 1637.
30. Yang, Y.C., Ward, J.R., Luteran, T. (1986) *J. Org. Chem.* **51**, 2756.
31. Yang, Y. C., Ward, J. R., Wilson, R. B., Burrows,W.,Winterle, J. S. (1987) *Thermochim. Acta* **114**, 313.
32. Yang, Y.C., Szafraniec, L.L., Beaudry, W.T., Ward, J. R. (1990) Report, CRDEC-TR-216.
33. Yang, Y.C., Ward, J.R., Hallowell, S.F.(1988) Report , CRDEC-TR-88043.
34. Yang, Y.C., Ward, J.R. (1988) Report, CRDEC-TR-88040.
35. Appler, B., Maschmeyer, D., Molsberger, D. (1983) *Untersuchungen zur Hydrolyse von 2,2'-Dichlorethylsulfid*, Wehrwissenschaftliche Dienststelle der Bundeswehr für ABC-Schutz, Munster, 1983.
36. Martinetz, D. (1992) *Z. Chem.* **22**, 256.
37. Martinetz, D., Lohs, K. (1985) *Z. Chem.* **25**, 400.
38. Martinetz, D. (1983) *Mitt. Chem. Ges. DDR* **30**, 226.
39. Lohs, K., Martinetz, D.(1978) *Entgiftungsmittel und Entgiftungsmethoden*, p. 117, Vieweg, Braunschweig.
40. Martinetz, D. (1981) Immobilisation, *Entgiftung und Zerstörung von Chemikalien* **202**, pp. 104, Verlag Harri Deutsch Thun, Leipzig.
41 Bales, S., Nickelson (1922) *J. Chem. Soc.* **121**, 2137.
42. Bales, S., Nickelson (1923) *J. Chem. Soc.* **123**, 2486.
43. Satori, M. (1939) *Die Chemie der Kampfstoffe*, p. 250, Vieweg, Berlin.
44. *ibid.* p. 248.
45. Jackson, K.E. (1934) *Chem. Rev.* **15**, 425.
46. Franke, S., Koehler, K.F., Zaddach, H. (1993) *Chemie der Kampfstoffe*, Teil 1, p. 264, Dr. Koehler mbH, Munster.
47. *ibid.* p. 242.
48. Tilley, R.I. (1993) *Aust. J. Chem.* **46**, 293.
49. Hanslian, R. (1937) *Der Chemische Krieg, Teil 1 Militärischer Teil*, Verlag von E. S. Mittler&Sohn, Berlin, p. 392.
50. Reichert, C., Lough, C.E. (1974) Study of 'mustard' destruction by reaction with aqueous sodium sulfide, *DRES-Suffield Technical Note*, No **342**.
51. Reichert, C. (1975) Study of 'mustard' destruction by hydrolysis, *DRES-Suffield Technical Note*, No **329**..
52. Reichert, C. (1975) Hydrolysis of 'mustard' in full-scale reaction vessel, *DRES-Suffield Technical Note,* No **352**.
53. Seiders, R.P., Ward, J.R. US Pat. Appl. 870,215 (March 03, 1987) [Chem. Abstr. 1987 106, 109950n].
54. Trofimov, B. A., Amosova, S. V., Gusarova, N. K., Kryuchkov, V. V., Efremova, G. G., Parygina, V.I., Babkin, V.A. (1975) *Tezisy Dokl. Vses. Konf. Khim. Atsetilena 5th* **58**.
55. Trofimov, B.A., Gusarova, N.K., Amosova, S.A., Efremova, G.G. (1979) *Otkrytiya, Izobret., Prom. Obraztsy, Tornaye Znaki* **32**, 236.
56. Glukharev, Y.A., Uzhdavini, E.R., Gilev, A.A., Stepanova, A.A. (1980) *Gig Tr. Prof. Zabol.* **7**, 51.

OLD 'MUSTARD' OR YPERITE MUNITIONS: METHODS FOR DESTRUCTION AND CONTAINER DETOXIFICATION

***Report of the Yperite Discussion Group*[1]**

Problems associated with the disposal of 'mustard' CW agent have not changed in the period since initiation of large scale destruction programs. In particular, the presence of materials other than the pure chemical compound, bis(2-chloroethyl) sulfide, in the CW agents greatly affects the suitability of processes that may be proposed for their destruction. These additional materials include additives (e.g., thickeners or other chemical agents such as Lewisite), by-products of the manufacturing process, and products that arise from aging of the CW agent.[1] Thus, the destruction of 'mustard' involves the conversion of a multitude of chemical species into material suitable for ultimate disposal. Importantly, the overall destruction of the 'mustard' CW agent also requires that the material be presented in the correct form to the reaction stage of the process (e.g., as a liquid or in solution for hydrolysis).

Given the wide range of materials that may contaminate old 'mustard', non-specific destruction technologies that can accommodate the full range of physical and chemical properties of the material have received the most attention by those tasked to destroy CW agent stockpiles. Currently, only high temperature oxidation (i.e., incineration) has been demonstrated to be suitable for large scale destruction of all the materials that are collectively referred to as 'mustard' CW agent. For a number of reasons, including the small scales of operations at some sites, and public opposition, incineration may not be appropriate or acceptable for the destruction of 'mustard' at some locations.

If incineration is excluded, the diversity of physical and chemical properties displayed by 'mustard' CW agent and the diverse scales of destruction processes at individual sites probably will necessitate that a number of different processes be available to achieve the goals of a destruction program. Indeed, there is heightened recognition that one process may not be suitable in all instances. Development of procedures for application to specific disposal problems should be undertaken. For instance, development of a process or processes for small scale destruction/conversion of liquid 'mustard', or 'mustard' containing additional material, to products suitable for immediate disposal or transport to a site for ultimate disposal. This work must be approached with full recognition that the perceived limitations of a method for destruction of 'mustard' CW agent in one set of circumstances may not be relevant in another.

Recent progress in hydrolysis followed by bioremediation provides an excellent

1Report of a Discussion Group Comprising Participants E. Koch, M. Koch, Leslie, Mikolajczyk, Modena, Mulbry, Petrosyan, Sokolowski and Yang.

J.F. Bunnett and M. Mikołajczyk (eds.), Arsenic and Old Mustard: Chemical Problems in the Destruction of Old Arsenical and 'Mustard' Munitions, 175–176.

Recent progress in hydrolysis followed by bioremediation provides an excellent example of a process developed for destruction of 'mustard' CW agent at a specific location.[2] In general, hydrolysis in water has not been considered a viable process for the destruction of 'mustard' due to the inability of hydrolysis to destroy additional materials present in aged or modified 'mustard' CW agent. Nevertheless, hydrolysis in a large excess of water (100 fold by volume) followed by bioremediation and then UV/peroxide treatment has progressed to large scale trials at the U.S. Army ERDEC center. It is expected that the process will be applicable to the destruction of material at one site that constitutes, however, only about 2% of the U.S. stockpile. In another recent development, electrochemical pretreatment of products from hydrolysis of 'mustard' and Lewisite (an additive to 'mustard' in some CW munitions) has been demonstrated to yield a waste stream compatible with bioremediation prior to final release or disposal of the material.

In this spirit, technologies that for one reason or another have not been pursued or have been disregarded should be reconsidered in light of the wide range of problems that exist for the destruction of 'mustard' CW agent.

Future research should address methods for degradation of aged 'mustard' that contains polymerised or other solid material. These should include electrochemical methods (such as the Silver II process, which can provide complete mineralisation, and electrolysis to selected stage of conversion), and chemical reduction processes such as reaction with sodium/solvated electrons,[3] hydrocracking and pyrolysis. These efforts should include extenstion of procedures that have been developed for destruction of well characterised 'mustard' CW agent (e.g., hydrolysis with or without electrochemical treatment followed by bioremediation) to 'mustard' containing additional, more complex materials.

The development of systems to disperse the polymerised and solid materials in old 'mustard' so as to make all the components accessible to chemical transformation remains a critical area for research. Investigations of new solvent systems and molecules with novel structures as dispersants (perhaps dendrimers or other materials with unusual surface properties), to facilitate the solubilisation and transfer of the CW agent, might provide major advances in the development of destruction procedures.

Research into the cleanup of contaminated sites is also necessary. This includes bioremediation of soil and chemical transformations in the soil. Site cleanup must deal with the chemical agent, the full range of additional material that may be present in CW agent, and products of partial destruction.

References.

1. Norman, P.R. *This volume,* p.105
2.. Harvey, S.P., Blades, T.A., Szafraniec, L.L., Beaudry, W.T., Haley, M.V., Rosso, T. and Young, G.P. *This volume,* p.115
3. Bunnett, J.F. *This volume,* p. 163

OLD ARSENICAL MUNITIONS: METHODS FOR DESTRUCTION AND SITE CLEANUP

***Report of the Arsenicals Discussion Group*[1]**

Some arsenical munitions are recovered entire. Others have spilled the arsenical agent owing to corrosion or external blows, and some arsenical agents were spilled on the earth in the course of filling, transporting or testing of munitions. These situations present problems of differing sorts.

The fact that several kinds of arsenic compounds are very poisonous complicates their disposal. Worst are probably arsine (AsH_3) and arsenic trioxide (As_2O_3). Arsenate salts have lower acute toxicity but appear to be carcinogens. Some organic compounds of arsenic have remarkably low toxicity, as noted in the lecture by Cullen (this volume).

The final arsenic-containing products of arsenicals destruction must somehow be returned to the earth. Inasmuch as the arsenic used to manufacture arsenical agents came from the earth, well-considered land burial of arsenic compounds of low-toxicity can be viewed as restoration of the natural situation.

1. Spilled arsenical agents

Salient problems are (1) how much chemical degradation has occurred from exposure to water, oxygen, sunlight and the like, (2) what sort of biotransformation may have occurred thru the action of organisms present at the site, (3) and how to decontaminate soil that contains arsenical agents or transformation products therefrom.

1.1 NATURAL CHEMICAL TRANSFORMATIONS *IN SITU*

The As–Cl bonds of arsenical agents are somewhat susceptible to hydrolysis, furnishing not As–OH functionality but rather dimers with As–O–As bridges. The lecture of Martens (this volume) tells of work to detoxify patches of arsenicals-contaminated soil at the former German chemical munitions filling plant near Munster.

[1]Report of a Discussion Group Comprising Participants Beletskaya, Bilger, Boronin, Bunnett, Costantino, Cullen, Dominas, Goessler, Haiduc, Maeda and Martens.

J.F. Bunnett and M. Mikołajczyk (eds.), Arsenic and Old Mustard: Chemical Problems in the Destruction of Old Arsenical and 'Mustard' Munitions, 177–183.

1.2 BIOTRANSFORMATIONS

Terrestrial plants as well as aquatic organisms can accumulate arsenic compounds. Algae in the sea do so, utilizing arsenic normally present in sea water, and aquatic organisms that graze upon such algae incorporate the arsenic in their tissues. (See lectures in this volume by Cullen and Maeda). Plants that are able to grow on arsenicals-contaminated soil may take up some of the arsenic compounds. [At a German site near Munster, *Polygonium sachalinense* has been grown successfully on soil heavily contaminated with arsenicals.] As a method for site remediation, that technology would conceivably be complicated by legal constraints on disposal of arsenic-containing plant material. If however the plant materials are harvested, dried and incinerated, the arsenic disposal problem is like that from incineration of arsenical agents.

Little is known about the growth of microorganisms in arsenic-contaminated soil. A frightening possibility is that the microorganisms would release arsenic in some volatile form, such as arsine. Even were that not to occur, biodegradation of arsenical chemical warfare agents would not be fully adequate for decontamination purposes, for arsenic compounds would remain in the soil.

1.3 DECONTAMINATION OF SOIL THAT CONTAINS ARSENICAL AGENTS

In situ chemical destruction of arsenicals is conceivable in principle, but there is little practical knowledge of it. For example, a contaminated site might be soaked with aqueous sodium sulfide, so as to obtain deep penetration; the sulfide anion, being highly reactive with As–Cl functionality, is expected to substitute sulfur for chlorine [1,2,3], forming substances of much reduced toxicity. Although such treatment might be unacceptable for the long term, immediate danger at the site would be reduced.

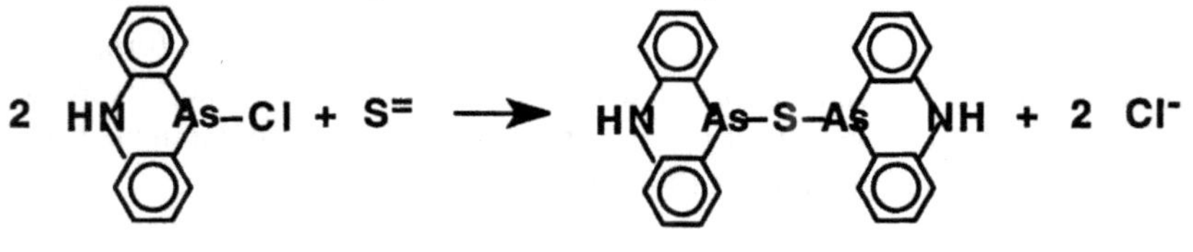

Expected reaction of Adamsite with sulfide ion

1.4 REMOVAL OF CONTAMINATED SOIL FOR TREATMENT ELSEWHERE

Even at a heavily-contaminated site, the mass of the arsenical agents is much less than of the whole soil. Processing all the contaminated soil involves handling tons of largely inert material. Also, when a thermal treatment is involved (e.g., incineration), heating all the inert material is energy-expensive. That problem is being simplified in Germany by, in essence, laundering the soil, as Martens reports (this volume). It is agitated in water with flotation agents, which release the toxic materials from sand and clay. The washed soil is returned to its place, and the toxic materials are incinerated.

2. Destruction of arsenical agents in containers (munitions or storage tanks)

From containers, liquid arsenical agents can be drained. The problem then is how best to destroy the agents. But, the emptied containers remain contaminated with agent, which must be destroyed before containers are placed in landfill or sold as scrap metal.

2.1 INCORPORATION INTO CONCRETE

In Italy, an unconventional method is being used to dispose of Adamsite: finely ground, it is mixed with Portland cement, sand and water; the concrete "mix" is poured into concrete boxes and allowed to set as concrete; for the purpose of physical isolation, the hardened concrete is covered by a layer of paraffin wax and then by a concrete cover.

In discussion, questions were raised as to the chemical form of Adamsite in concrete; has the substance undergone decomposition, for instance, As–Cl being converted to As–OH functionality in the strongly alkaline medium? If so, has that product further reacted so as to become chemically bonded into the complex inorganic networks of concrete? Or could Adamsite retain its molecular structure and somehow escape from the concrete matrix, despite its low solubility and volatility?

2.2 INCINERATION

The arsenical chemical warfare agents may be incinerated, usually admixed with fuel oil. The arsenic constituent usually is converted to As_2O_3, which must be removed from the flue gases because of its high toxicity. It is often "scrubbed" from those gases by means of aqueous sodium hydroxide, in which it is converted to sodium arsenite (Na_3AsO_3), also lethal. In German practice, the sodium arsenite is oxidized to arsenate (Na_3AsO_4). Addition of an iron (III) salt to a sodium arsenate solution precipitates iron (III) arsenate ($FeAsO_4$), which has very low solubility in water,. It can be collected by filtration and ultimately stored in a remote location, such an abandoned, deep mine.

Incineration is furthermore applicable to destruction of mixtures of an arsenical agent with another compound, such as another agent, or a polymer added as thickener. In particular, the vesicant Lewisite has sometimes been mixed with yperite or 'mustard'.

An advantage of incineration is that it serves to clean out residues of agent in emptied containers. A special container decontamination furnace may be employed.

2.3 REACTION WITH ALKALIES

The reaction of Lewisite with sodium hydroxide is well known:

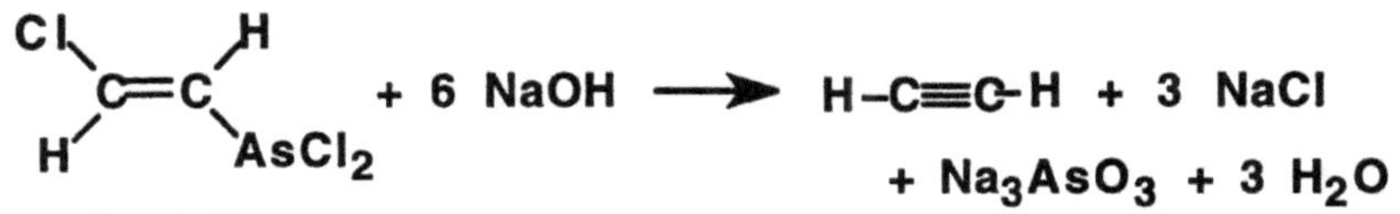

Lewisite

The sodium arsenite product is an article of commerce, although of low value. Petrov et al. (this volume) describe the analogous reaction with sodium alkoxides, which forms trialkyl arsenite esters instead of sodium arsenite.

Clark I and Adamsite both react with alkali, but rather than Ph_2AsOH or its analog from Adamsite, dimers of type $Ph_2As–O–AsPh_2$ are obtained [3,4,5].

2.4 CLEAVAGE BY SODIUM

Sodium metal aggressively attacks many compounds, especially those with bonds to chlorine or other halogen, often to tear the molecules apart. All the arsenical agents are expected to react with sodium metal; for example, in the case of Clark I:

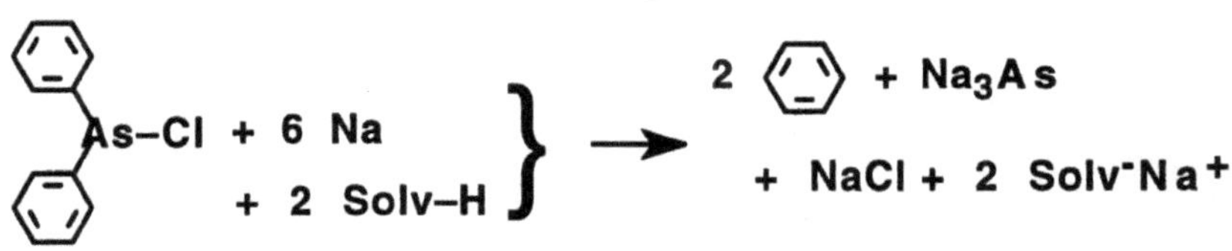

Reaction of Clark I with Sodium

("SolvH" represents a hydron donor such as water, ammonia or an alcohol, and "Solv-Na^+" its sodium salt.) Attractive features of the use of sodium to destroy chemical warfare agents are mentioned in the poster abstracts of Sokolowski and Bilger and of Bunnett (this volume). A potential problem is that adventitious acids that might somehow mix with the products will convert sodium arsenide (Na_3As) to the extremely toxic arsine (AsH_3). Conversion of Na_3As to a much less toxic arsenic compound should immediately follow the sodium treatment.

2.5 HYDROGENOLYSIS.

Like most other chemical warfare agents, the arsenicals can be degraded by hydrogenolysis (Petrov et al., this volume). The arsenic is liberated as arsine. Because of the extreme toxicity of this gaseous substance, many nations are likely to impose stringent requirements for capture of it. That could complicate engineering of the process, and pose problems in getting permits from authorities.

2.6 NOVEL METHODOLOGIES FROM RUSSIA.

Petrov et al. (this volume) describe novel reactions of Lewisite with elemental sulphur and with the mono-acrylic ester of ethylene glycol to form polymers, as well as destruction of Lewisite by reaction with elemental chlorine or with ammonia.

The Russian chemical munitions stockpile contains appreciable amounts of mixtures of Lewisite with yperite. The mentioned polymerization processes show promise also for destruction of those agent mixtures.

2.7 OXIDATION IN WATER SOLUTIONS.

In wet air oxidation [6], a rather dilute (say, 1-10%) solution or suspension of an organic waste in alkaline water is heated to a temperature somewhat below the critical temperature of water (374°C), in the presence of an oxidizing agent, which is sometimes hydrogen peroxide and sometimes dioxygen. Oxidation of the organic components occurs rapidly. It is expected that arsenical agents will be extensively oxidized, the arsenic converted to inorganic forms, and organic residues degraded to carbonate as well as low molecular weight compounds such as acetic acid. Further treatment to capture the arsenical products would be necessary.

2.8 OXIDATION UNDER ULTRAVIOLET IRRADIATION.

An effective method for destruction of other organic wastes is ultraviolet irradiation of a water solution or slurry containing also hydrogen peroxide. It should effectively destroy arsenicals or products therefrom insofar as they can be brought into solution in water. Capture of the arsenical oxidation products would be necessary before any release into the environment.

In such procedures, the aggressive species chiefly responsible for attack on pollutants is hydroxyl radical (HO•). Hydroxyl radicals can also be generated by irradiation of water solutions of organic wastes that contain colloidal semiconductor particles, such as zinc oxide and titanium dioxide [7,8]. Sometimes sunshine can serve as the source of irradiation. In closely allied procedures, the hydrogen peroxide is accompanied by a peroxy-isulfate ($S_2O_8^{=}$) salt. The peroxydisulfate salt may purchased or, in a variation due to Cooper [9], generated by electrolysis of sulfate salts in water.

2.9 SILVER(II) OXIDATION.

A method advocated for destruction of most chemical warfare agents is oxidation by silver ions in the unusual +2 oxidation state [10,11]. Agent destruction occurs within the electrochemical cell for generation of the silver(II). If this process were applied to destruction of arsenicals, the arsenic would undoubtedly be oxidized to arsenate (AsO_4^{3-}) which could be captured by addition of iron(III) sulfate to precipitate $FeAsO_4$.

2.10 FURTHER TECHNOLOGIES.

Reports in this volume by Chimishkyan (p. 155) and by Wertejuk et al. (p. 91) describe methods for destruction of Adamsite developed in Russia and Poland, respectively. For arsenical agents of all types, other possibilities include oxidation of the arsenical agent with sodium hypochlorite (NaOCl) in aqueous medium, and oxidation by a mixture of molten salts that includes oxidants such a sodium nitrate and sodium peroxide.

2.11 BIODEGRADATION.

Although microorganisms that can degrade arsenical agents doubtless exist, biodegradation does not seem attractive for their destruction, unless in a closed system. Products of biodegradation would necessarily include arsenic compounds, some of which (e.g., arsine) might escape if biodegradation were conducted in open vats.

3. Need for research

The destruction of arsenical agents appears to have received little attention thru research. One factor is that arsenical agents are at most minor components in the chemical weapons stockpile of the United States, which has therefore not invested much in research on them. It is probable that research would reveal features of the chemistry of arsenical agents that would facilitate the design of effective technologies for their destruction.

Because arsenical munitions that have lain in the earth for decades are often heavily corroded, too fragile to survive transport to a central location, it is desirable to be able to destroy them at the sites where they are unearthed. That implies a need for portable equipment engineered to avoid agent release during the destruction process. Development of such equipment is desirable. A portable reactor might be suitable for destruction of many agents (by incineration or alternative technology) or it might need to be specially engineered for each technology.

4. References

1. Doak, G. O. and Freedman, L. D. (1970) *Organometallic Compounds of Arsenic, Antimony and Bismuth*, John Wiley & Sons, Inc., New York; p. 79.
2. Haiduc, I. (1970) *The Chemistry of Inorganic Ring Systems*, John Wiley & Sons, Inc., New York; Vol. 2, p. 894.
3. Wieland, H. and Rheinheimer, W. (1921) Über ringförmige Arsenverbindungen der Aromatische Reihe, *Justus Liebigs Ann. Chem.* 423, 1-38.
4. McKenzie, A. and Wood, J. K. (1920) Some organic compounds of arsenic,. *J. Chem. Soc.* 117, 406-415.
5. Pope, W. J. and Turner, E. E. (1920) Triphenylarsine and diphenylarsenious salts, *J. Chem. Soc.* 117, 1447-1452.
6. Mishra, V. S.; Mahajani, V. V.; Joshi, J. B. (1995) Wet air oxidation, *Ind. Eng. Chem. Res.* 34, 2-48.
7. Grätzel, C. K.; Jirousek, M.; Grätzel, M. (1990) Decomposition of Organo-phosphorus Compounds on Photoactivated TiO_2, *J. Mol. Catalysis* 60, 375-387.
8. Mabury, S. A.; Crosby, D. G. (1996) Pesticide Reactivity toward Hydroxyl and its Relationship to Field Persistence *J. Agric. Food Chem.* 44, 1920-1924.
9. Cooper, J. F. (1992) Destruction of Chemical Warfare Agents by Oxidation using Catalyzed Peroxydisulfate and Peroxide in a Closed System, unpublished manuscript (Lawrence Livermore Laboratory, California).
10. Farmer, J.C.; Wang, F.T.; Hawley-Fedder, R.A.; Lewis, P.R.; Summers, L.J.; Foiles, L. (1992) *J. Electrochem. Soc.* 139, 654-662.
11. R. S. Magee et al. (1996) Review and Evaluation of Alternative Chemical Disposal Technologies, U.S. National Research Council, Washington, DC.

ON THE HANDLING OF RECOVERED MUNITIONS

Report of the Recovered Munitions Discussion Group[1]

1. Activities on Site

After old munitions are located, they will have to be identified prior to any further activities taking place. Comments on this aspect are made in II. below. Accurate identification is essential for all future activities.

A value risk assessment must be made of the options available at each site, on a site by site basis. Likely options are either to treat the item on site or to transport the item to a central facility for remediation

If munitions are not to be moved (for whatever reasons), they must be disposed of *in situ.* Currently, this is achieved by open air detonation, but this option is likely to become unacceptable. The solution will be to develop new technologies for destruction that are transportable and can be used in the field. These technologies will also need to be backed up by a portable containment facility that will provide both chemical and ballistic integrity should an unexpected release or detonation occur.

2. Identification

Further research and development on technologies for accurate identification will be required.Although some useful methods are in hand, additional effort is perceived as highly desirable, particularly in the development of portable, low cost devices such as those based on acoustic methods.

2.1. NON INTRUSIVE METHODS

This is a critical area since accurate identification of a munition is essential to ensure that the correct procedures for rendering it safe are applied. Since it is likely that on site remediation will be the preferred option for dealing with small numbers of old and/or abandoned munitions, correct identification will be required to ensure that correct safety procedures are put in place. Although much progress has been made in recent years, the area will still benefit from intensive research. Current methods include:

1 Report of a Discussion Group Comprising Participants Blades, Froment, Holm, Malecki, Manley, Norman, Santesson, Soilleux, Stock, Zavialova and Zoubrilin.

J.F. Bunnett and M. Mikołajczyk (eds.), Arsenic and Old Mustard: Chemical Problems in the Destruction of Old Arsenical and 'Mustard' Munitions, 185–188.

(a) Cleaning and visual inspection - skilled technicians can learn a lot about a munition from its appearance and (with luck) its markings.

(b) X-ray - this is deemed to be absolutely essential for ascertaining the safety of recovered munitions. The technology is available, tried and tested, but needs to be made more user-friendly.

(c) Neutron methods: Data from first generation equipment is encouraging. It is going to be the only technology that will provide the elemental composition of weapon fills.

(d) Acoustic methods may be able to provide information on the physical form of a munition fill. Development in this area has been less successful than neutron methods. The techniques, if they can be developed, would provide cheap and portable instrumentation for field use.

2.2. INTRUSIVE METHODS

Sampling and analysis will almost always be required. Though the principles are well established, there is a need for a system more self reliant and relatively easy to use, suitable for use in the field.

3. Transportation

Standards for packaging, labelling etc. are well established in some countries. Once identified and packaged according to regulations, transport of alchemical munition by an appropriate carrier should be relatively straightforward.

(a) Bulk CW agents or contaminated soil may require transportation to a central destruction facility. To achieve this, international freight regulations for toxic material will need to be considered.

(b) Munitions will require a decision on safety. Explosively configured? Fuzed? Fired? CW agent? Explosive content? Stability? Suitable containers of several types are available in several countries, including containers capable of withstanding small explosions. These are however extortionately expensive and difficult to move.

On site remediation of soil is an attractive option, but will require a better understanding of the long term fate and effects on the environment of CW agents, their degradation products and remediation reagents.

4. Recovery from the Sea

This is not seen as a high priority. Highest priority will be cleaning up land based CW stockpiles and old munitions; recovery from the sea will initially have a low priority. Also, the little evidence that exists (c.f. Helsinki Commission CHEMU) would suggest that, with few exceptions, it is probably safer to leave sea dumped munitions where they are rather than attempt recovery. The exceptions are likely to be when CW munitions are found on beaches and might come into contact with people. Accurate predictions of the environmental impact of such events are difficult to make because of the paucity of data on both the long and short term fate and effects of such materials in a maritime environment. Research to address this omission is strongly recommended

Recovery from beaches is relatively straightforward and takes place regularly. The main problem associated with this activity is accurate identification, prior to rendering the munitions safe. Identification is made difficult by encrustation with marine organisms to a depth of several centimetres. Munition cleaning is a general problem, but this is an extreme example. Development of effective ways of cleaning items would be of great benefit in both land and sea recoveries of CW munitions.

5. Recommendations

The group recommends (as drawn from the above text) the following:

(a) Development of alternate technologies for *in situ* remediation of CW. Ideally such technologies should be easily sustainable in the field and of relatively low cost.

(b) Study of both the long and short term fate and effects of CW agents in marine environments.

(c) Research and development into low cost NDE equipment to aid screening and identification of CW munitions. Acoustic methods offer potential in this aspect.

(d) Collection of data on the long term environmental effects of *in situ* soil remediation methods. These data are needed to enable all criteria to be considered in preliminary risk assessments.

(e) Work to develop portable facilities to contain CW agents as well as ballistic effects.

(f) Consultation of available radiographic technology to derive a more user friendly radiographic system for field use. The group consider radiography to be absolutely essential for safe operation of all recovery and remediation procedures.

(g) Development of sampling and analytical facilities more suited to field use.

(h) Identification of technologies for adequate cleaning of recovered items.

In recommending all the above, the group is aware that some answers will already exist in other technology areas. Efforts should be made to gain access to these alternate areas in order to avoid reinventing the wheel.

APPENDIX A: PARTICIPANTS

Professor Irina P. Beletskaya
Moscow State University
Moscow 119899, RUSSIA

Dr. Edgar Bilger
Dr. Bilger Umweltconsulting GmbH
D-63579 Freigericht, GERMANY

Dr. Timothy Blades
U. S. Army ERDEC
Aberdeen Proving Ground, MD

Dr. Alexander Boronin
Pushchino Biological Research Center
Pushchino 142292, RUSSIA

Dr. Bogdan Bujnicki
Centre of Molecular and Macromolecular Studies
90-363 Lodz, POLAND

Professor Joseph F. Bunnett
University of California
Santa Cruz, CA 95064, USA

Professor A. L. Chimishkyan
D. Mendeleev University of Chemical Technology
Moscow 125190, RUSSIA

Lt. Col. Giacinto Costantino
Army Technical Corps Headquarters
Rome, ITALY

Professor W. R. Cullen
University of British Columbia
Vancouver, B.C., CANADA

Col. Dr. Wojciech Dominas
Military Institute of Chemistry and Radiometry
00910 Warszawa, POLAND

Professor Josef Drabowicz
Centre of Molecular and Macromolecular Studies
90-363 Lodz, POLAND

Monsieur Daniel Froment
Délégation Générale pour l'Armement
91710 Vert-le-Petit, FRANCE

Mag. Walter Goessler
Karl-Franzens University
8010 Graz, AUSTRIA

Col. Dr. Stanislaw Gruchalski
NBC Defence Corps
Polish Army Central Staff
00-909 Warszawa, POLAND

Dr. Krzysztof Guttler
Centre of Molecular and Macromolecular Studies
90-363 Lodz, POLAND

Dr. Ionel Haiduc
Babes-Bolyai University
RO-3400 Cluj-Napoca, ROMANIA

J.F. Bunnett and M. Mikołajczyk (eds.), Arsenic and Old Mustard: Chemical Problems in the Destruction of Old Arsenical and 'Mustard' Munitions, 189–192.

Dr. Francis W. Holm
S. A. I. C.
Aberdeen, MD 21001, USA

Ms. Kim Jackson
University of California
Santa Cruz, CA 95064, USA

Professor Frantisek Kastanek
Institute of Chemical Process
Fundamentals
Czech Academy of Sciences
16502 Prague, CZECH REPUBLIC

Dr. Piotr Kielbasinski
Centre of Molecular and Macromolecular
Studies
90-363 Lodz, POLAND

Dr. Ernst-Christian Koch
ABC und Selbstschutzschule
Sonthofen, GERMANY

Dr. Mieczyslaw Koch
Military Institute
of Chemistry & Radiometry
00910 Warszawa, POLAND

Mr. Jerzy Krysiak
Centre of Molecular and Macromolecular
Studies
90-363 Lodz, POLAND

Dr. Ralph Leslie
Defence Science & Technology
Organisation
Melbourne, VIC 3001, AUSTRALIA

Dr. Jerzy Luczak
Centre of Molecular and Macromolecular
Studies
90-363 Lodz, POLAND

Dr. Piotr Lyzwa
Centre of Molecular and Macromolecular
Studies
90-363 Lodz, POLAND

Professor Shigeru Maeda
Kagoshima University
Kagoshima 890, JAPAN

Col. Marek Malecki
Ministry of National Defence
00-911Warszawa, POLAND

Mr. Ron Manley
Chemical Weapons Branch, OPCW
Den Haag, NETHERLANDS

Dr. Hermann Martens
Wehrwissenschaftliches Institut für
Schutztechnologien, ABC Schutz
29623 Munster, GERMANY

Dr. Wanda Midura
Centre of Molecular and Macromolecular Studies
90-363 Lodz, POLAND

Professor Marian Mikolajczyk
Centre of Molecular and Macromolecular Studies
90-363 Lodz, POLAND

Professor Giorgio Modena
Istituto di Chemica Organica dell' Universitá
35131 Padova, ITALY

Dr. Walter Mulbry
U.S. Agricultural Research Service
Beltsville, MD 20705; USA

Dr. Paul R. Norman
CBDE, Porton Down
Salisbury, Wilts. SP4 0JQ, UK

Professor Valery S. Petrosyan
Moscow State University
Moscow 119899, RUSSIA

Dr. Johan Santesson
Organisation for Prohibition of Chemical Weapons
Den Haag, NETHERLANDS

Professor Kurt Schaffner
Max-Planck-Institut für Strahlenchemie
Mülheim a.d. Ruhr, GERMANY

Ms. Nancy Schulte
NATO, Scientific and Environmental Affairs Division
Bruxelles, Belgium

Col. Prof. Boguslaw Smólski
Ministry of National Defence
00-911Warszawa, POLAND

Ms. Agata Sobczak
Centre of Molecular and Macromolecular Studies
90-363 Lodz, POLAND

Dr. Richard Soilleux
CBDE, Porton Down
Salisbury, Wilts SP4 0JQ, UK

Dr. M. Sokolowski
Military Institute of Chemistry & Radiometry
00910 Warszawa, POLAND

Dr. Ivan Starovoitov
Laboratory of Metabolism of Xenobiotics
Russian Academy of Sciences
Moscow Region, RUSSIA

Dr. Thomas Stock
SIPRI, Frösunda
S-171 53 Solna, SWEDEN

Col. Dr. Wiktor Tyszkiewicz
Military Institute
of Chemistry & Radiometry
00910 Warszawa, POLAND

Dr. Zbigniew Wertejuk
Military Institute
of Chemistry & Radiometry
00910 Warszawa, POLAND

Dr. Andreas Wimmer
Bundesministerium für Verteidigung
A-1024 Wien, AUSTRIA

Dr. Yu-Chu Yang
U.S. Army ERDEC
Aberdeen Proving Ground, MD, USA

Dr. Nataliya Zavialova
Scientific Centre of Chemical Forces
Ministry of Defence
Moscow, RUSSIA

Professor Vadim P. Zoubrilin
Scientific Centre of Chemical Forces
Ministry of Defence
Moscow, RUSSIA

APPENDIX B: MEMBERS OF THE IUPAC COMMITTEE ON CHEMICAL WEAPONS DESTRUCTION TECHNOLOGIES

Profesor Wataru Ando
Department of Chemistry
University of Tsukuba
Tsukuba, Ibaraki 305, JAPAN

Professor I. P. Beletskaya
Chemistry Department
Moscow State University
Moscow 119899, RUSSIA

Professor Joseph F. Bunnett
Department of Chemistry & Biochemistry
University of California
Santa Cruz, CA 95064, USA

Dr. Dupont Durst
Research Directorate
U.S. Army ERDEC
Aberdeen Proving Ground, MD 21010, USA

Dr. Claude Eon
Ministére de la Défense
Centre d'Études du Bouchet
BP n° 3
91710 Vert-le-Petit, FRANCE

Dr. Daniel Froment
Ministére de la Défense
Centre d'Études du Bouchet
BP n° 3
91710 Vert-le-Petit, FRANCE

Dr. Ralph Leslie
Defence Science and Technology Organisation
P.O. Box 4331
Melbourne VIC 3001
AUSTRALIA

Mr. Ronald G. Manley
Preparatory Commission for OPCW
Laan den Meerdervoort 51
Den Haag, The Netherlands

Dr. Blaine C. McKusick
1212 Bruce Road, Carrcroft
Wilmington, Delaware 19803-4202

Professor Marian Mikolajczyk
Centre of Molecular and Macromolecular Studies
Polish Academy of Sciences
Sienkiewicza 112
90-363 Lód´z, POLAND

Professor Giorgio Modena
Dipartimento di Chimica Organica
Universitá degli Studi di Padova
Via Marzolo, 1
I-35131 Padova, ITALY

J.F. Bunnett and M. Mikołajczyk (eds.), Arsenic and Old Mustard: Chemical Problems in the Destruction of Old Arsenical and 'Mustard' Munitions, 193–194.

Dr. Walter Mulbry
Pesticide Degradation Laboratory
US ASRR
Room 100, Bldg. 050, Barc-West
Beltsville, MD 20705

Dr. Graham S. Pearson CB
Fieldway, Winterbourne Monkton
Swindon, Wiltshire SN4 9NW
UNITED KINGDOM

Professor Dr. Kurt Schaffner
Max-Planck Institut für
Strahlenchemie
Stiftstrasse 34-36
D-4330 Mühlheim a. d. Ruhr 1
GERMANY

APPENDIX C: ON WHOLESOME WARFARE

Chemical warfare as we know it today was initiated in 1915. It is generally disapproved. Adjectives such as *abhorrent*, *barbarous*, *dreadful*, *gruesome*, *horrible*, *inhumane* and *sinister* have frequently been applied to it.

For perspective, let us view an episode of warfare as practiced before 1915. We quote from the book *Tsushima*, by A. Novikoff-Priboy, who was a sailor on a Russian naval vessel involved in battle with Japanese naval forces in 1905.[1]

"Boatswain Voevodin, having succeeded in extinguishing a fire which had broken out in the paint-room, went forward. The torpedo-hand Vasya Drozd came to meet him on the upper deck. Vasya walked swiftly, though stooping, holding one hand over the top of his head, while waving the other as if it could ward off shells. Whither away in such a hurry, this lean, long-legged dreamer? Anyhow, he stopped short in his stride, to look at the enemy. At this precise moment the boatswain was knocked down by the wind of a shell. He rose, uninjured, and caught sight of something which, surrounded by grey smoke, was jumping like a young bear. When the smoke had blown away, the boatswain could hardly believe his eyes. Vasya Drozd, with both legs shot off at the knee, was in the death-struggle. His eyes anguished, his face twisted, he was bounding on the red stumps that were left him, trying (it seemed) to run, and fighting desperately for breath. A moment later he fell into the pool of his own blood.

"'Brothers, the ironclad is flying up into the clouds, up into the clouds,' he cried, in a heart-rending voice.

"He continued to roll on the deck, this fragment of what had just before been a man, now laughing savagely, now uttering hysterical shrieks. Then the noise was stilled. The death-agony was over."

---

1 A. Novikoff-Priboy (1937), *Tsushima*, Alfred A. Knopf, New York, p. 173.

J.F. Bunnett and M. Mikołajczyk (eds.), Arsenic and Old Mustard: Chemical Problems in the Destruction of Old Arsenical and 'Mustard' Munitions, 195.

Author Index

J.F. Bunnett and M. Mikołajczyk (eds.), Arsenic and Old Mustard: Chemical Problems in the Destruction of Old Arsenical and 'Mustard' Munitions, 196.

Subject Index

J.F. Bunnett and M. Mikołajczyk (eds.), Arsenic and Old Mustard: Chemical Problems in the Destruction of Old Arsenical and 'Mustard' Munitions. 197–200.

MIX
Papier aus verantwortungsvollen Quellen
Paper from responsible sources
FSC® C105338

If you have any concerns about our products,
you can contact us on
ProductSafety@springernature.com

In case Publisher is established outside the EU,
the EU authorized representative is:
Springer Nature Customer Service Center GmbH
Europaplatz 3, 69115 Heidelberg, Germany

Printed by Libri Plureos GmbH
in Hamburg, Germany